D1015449

Canine and Feline
Behavioral Therapy

Canine and Feline Behavioral Therapy

BENJAMIN L. HART, D.V.M., Ph.D.
and
LYNETTE A. HART, Ph.D.
University of California, Davis

Lea & Febiger • Philadelphia

Lea & Febiger
600 Washington Square
Philadelphia, PA 19106-4198
U.S.A.
(215) 922-1330

Library of Congress Cataloging in Publication Data

Hart, Benjamin L.
Canine and feline behavioral therapy.

Includes index.
1. Dogs—Behavior. 2. Cats—Behavior. 3. Animal behavior therapy. I. Hart, Lynette A. II. Title.
SF433.H36 1985 636.7′089689 84-14401
ISBN 0-8121-0949-X

Copyright © 1985 by Lea & Febiger. Copyright under the International Copyright Union. All rights reserved. This book is protected by copyright. *No part of it may be reproduced in any manner or by any means without written permission from the publishers.*

PRINTED IN THE UNITED STATES OF AMERICA

Print No. 4 3 2

To Fern,
an inspiring mother

Preface

Behavioral problems in dogs and cats are so common that it is perhaps unusual to have a pet with no problems. Although people widely recognize the existence of pet behavioral problems, they are generally perplexed about how to proceed in correcting them. Pet owners are increasingly turning to professionals for assistance and advice. The small animal veterinarian is naturally one person who pet owners expect can help with the diagnosis and treatment of problem behavior. Another such professional is the animal behavior therapist who works in conjunction with a small animal hospital or on a referral basis with several hospitals.

This book is written first as a reference handbook, and secondly as a source of background information for those who are involved with behavioral problems in dogs and cats. We trust this book will find wide acceptance in the curriculum of schools and colleges of veterinary medicine as animal behavioral therapy becomes an established discipline. We have found that, along with concerned pet owners, both students and established practitioners eagerly welcome this relatively new area of animal health care.

This book is organized into three sections to facilitate efficient retrieval of necessary information. Section 1 covers practical tips in handling clients and in using the office visit to produce results. Section 2 comprises the crux of the book with eighteen chapters on specific behavioral problems. Throughout this section there are sample cases with brief instructions to clients for the most typical problems. We envision the readers making direct use of the 43 sample instructions by adapting them to the cases each practitioner sees. The chapters in this section can be referred to when a practitioner is confronted with a particular problem. Background material that one may want to read at his or her leisure comprises Section 3. Understanding the concepts presented in Section 3 will greatly facilitate the reader's effective use of the specific therapeutic approaches of Section 2.

A couple of people should be recognized for their foresight in encouraging the development of the area of dog and cat behavioral therapy. Ms. Nancy Bull, President of Veterinary Practice Publishing Company of Santa Barbara, California, recognized the critical importance of behavioral problems over a decade ago and provided visibility for the field in the journals, *Feline Practice* and *Canine Practice*. Literature on feline and canine behavior that has appeared in these journals over the years is available in monograph form from the publisher. Second, we appreciate the efficient and informative support provided by our editor at Lea & Febiger, George Mundorff.

Davis, California

Benjamin L. Hart
Lynette A. Hart

Contents

Section I

THE MEDICAL INTERVIEW AND MANAGEMENT OF BEHAVIORAL PROBLEMS

Section II

CAUSES AND THERAPEUTIC APPROACHES FOR COMMON BEHAVIORAL PROBLEMS

Section III

SCIENTIFIC BASES FOR TECHNIQUES OF BEHAVIORAL THERAPY

Section I

THE MEDICAL INTERVIEW AND MANAGEMENT OF BEHAVIORAL PROBLEMS

Behavioral problems with pets are familiar to all of us. Within any given week, a practitioner will hear references to an aggressive dog, a cat that sprays urine, or a dog that runs away. He or she will also treat medical problems that result from these behavioral problems. Fighting cats get infected lacerations, and roaming dogs come home with various wounds and injuries. Although no medical injuries usually result when cats spray, human tempers certainly flare.

Behavioral problems are prevalent in dogs and cats, and practitioners are regularly faced with treating them. We have all seen cases in which a behavioral problem is the primary disruptive influence in family dynamics, even more so than an illness of the pet. Such a problem merits the central attention of the practitioner.

What we set forth here is a systematic approach to assist the practitioner in handling behavioral problems, so that they can be dealt with more efficiently and effectively. In this section we lay some groundwork, conveying the attitude and interviewing style that one will find useful in working with problem behavior.

In the first chapter we overview the services included within the field of behavioral therapy, such as treating behavioral problems, evaluating problems that might be avoided in the future, and assisting in pet selection. In assessing a behavioral problem, the practitioner considers the pet, the owner, and the interactions between the pet and the owner. The organization of the book, also described in the first chapter, enables the practitioner to focus on specific treatment techniques (Section II) or on background information (Section III).

The second chapter examines the interview with the owner and pet, an occasion we view as the keystone of the therapeutic process. For a successful outcome, it is often necessary that the behavior of the owner and that of the pet change after only one or two brief sessions. This is a challenging goal, but one that is rewarding and at times easily achieved. In the interview, the practitioner can bring together various aspects of his expertise to affect an improvement in the problem.

Chapter 1

Introduction to Behavioral Therapy

Behavior is perhaps the most rapidly emerging, if not the most interesting, discipline in animal health care. Increasingly, dog and cat owners are seeking information about the normal behavior of cats and dogs and requesting help in dealing with behavioral problems. To some extent, this parallels the increasing sophistication of animal owners concerning the science of animal behavior, an interest stimulated by popular books, movies, and television specials on wild animals in various parts of the world. Some years ago small animal veterinarians reported that at least 3 to 4% of all cases represented primarily behavioral problems (Hart, 1971). Books on animal behavior frequently refer to behavioral problems, and often include examples of typical cases (Beaver, 1980; Hart, 1985; Houpt and Wolski, 1982). In dogs, the most frequent problem is aggressive behavior; in cats, urine spraying. Other common problems are changes in appetite, objectionable sexual behavior, phobias, destruction of household items, and roaming.

We do not have answers to all of these problems, and certainly we could improve upon some of our current recommended therapeutic approaches. Unfortunately, euthanasia has been one of the frequent answers to serious behavioral problems. Aside from the unnecessary loss of life of a pet, to perform euthanasia for a behavioral problem is undoubtedly more emotionally traumatic to pet owners than performing euthanasia for medical reasons such as a terminal illness. We sometimes have to remind our colleagues in the veterinary medical specialties that to save a pet from the fate of euthanasia by resolving a behavioral problem is as meaningful to the pet owner as saving the animal's life by the diligent use of clinical tests or the performance of skillful surgical operations.

THE GROWING FIELD OF BEHAVIORAL THERAPY

Behavioral therapy is not a highly technical field suitable for only a handful of specialists. Small animal practice is reaching the point where most practitioners are expected to offer help in the behavioral area. Like any discipline in which there are various levels of expertise, different practitioners will find the level at which they feel comfortable. Most of the chapters in this volume offer advice applicable to any small animal practice, although some aspects of behavioral therapy may be more appropriate for those who wish to offer behavioral treatment as a referral specialty.

Our services fall into three categories. One includes treating behavioral problems by administering drugs, instructing the client on interacting with the pet differently than before, or prescribing a detailed conditioning program. A second category of service is offering advice in identifying causes of specific problems that have occurred, even if the problem requires that we perform euthanasia. The advice may help the client to prevent a similar problem with the next pet. Thirdly, we can assist clients in deciding where and when they should obtain a pet, and also provide information concerning the techniques for housetraining and the behavioral effects of having a pet spayed or neutered.

Despite the image sometimes projected by popular writers, behavioral therapy should not be thought of as the art of dealing with overindulged, pampered pets. The concepts and principles of behavioral therapy are useful for treating serious problems that occur in relatively normal families. Correcting a behavioral problem is often the only alternative to euthanasia.

Behavioral therapy differs from other veterinary and medical specialities in that the practitioner must astutely assess the client, the pet, the problem, and their interaction. The ideal treatment of the pet may require an extensive effort by the client at changing the behavior. If the client's commitment appears insufficient for a particular treatment, the veterinarian must select an option for treating the problem that is more appropriate for the specific client. Thus, in behavioral therapy, understanding the client is as important as understanding the behavioral problem.

Understanding Behavioral Problems

At the outset the practitioner will want to assess the circumstances of the pet and its owner. What are the unique aspects of this particular situation that may play a role in the problem and its solution?

The Pet and Its Context. Most clinical problems involve behavior that is normal from the animal's standpoint but objectionable to people. We require pets to live in our homes and adapt to our human culture and daily schedules. While domestication has increased the compliance and adaptability of dogs and cats, our requirements sometimes exceed the tolerance of the pet.

Objectionable behavior may simply arise from a specific living situation. For example, the dog who barks excessively in a tiny apartment might not bark if housed on a small farm. It is not appropriate to refer to behavioral problems of these types as abnormal. It is best to reserve the term "abnormal" for behavioral patterns that are actually maladaptive to the animal and serve no purpose, even for animals in the wild. Idiopathic aggressive behavior, flank-sucking, and hyperkinesis in dogs and wool-chewing in Siamese cats are examples of abnormal behavior. Behavioral changes that are secondary to disease processes in one or more organ systems should also be considered abnormal.

Before instructing a client in a conditioning program or offering other types of advice, it may be useful to explore the possibility of eliminating a problem by changing some aspect of pet care or a household situation. Letting a dog stay in the house while the owners are gone may be a simple way to eliminate destructive behavior in the backyard. The owners may be reluctant to allow the dog in the house because of shedding. The most expedient way to resolve this behavioral problem, therefore, may be to deal with the shedding rather than attempting to eliminate the destructive behavior outside during the day.

Trivial details are sometimes quite important. For example, owners may find it almost impossible to act indifferently to a dog in their attempt to extinguish an attention-getting behavior such as snapping at imaginary flies because of the constant noise created by the jingling dog tags. Taping the tags together to eliminate this irritation can mean the difference between success or failure in resolving the problem.

The Owner. Many people who own dogs have a companionable attachment to their pets. The intensity of this attachment varies, from one characterized by a great deal of shared affection to a fraternal relationship with less psychological commitment. It is estimated that this companionable type of relationship comprises about 50% of owner-dog relationships. About 20% of dog owners seem to keep dogs primarily as status symbols or valued possessions (instead of collecting old cars or guns, they collect and show dogs). The remaining 30% view dogs as a nuisance or worry about the harm they may do (Wilbur, 1976).

The attachment of people to cats also varies. It is estimated that almost 60% of cat owners have little involvement with their pets; in fact, the need for less care and attention seems to be one reason they have cats rather than dogs. About 20% of cat owners have strong emotional ties with their pets, depending upon them for companionship the same as others do with dogs. As with dogs, about 20% of all cat owners view their animals primarily as possessions or status symbols (Wilbur, 1976).

Having some idea of the type of attachment a client has with his pet gives you information about the depth to which you should explore some behavioral problems. Owners with a strong emotional tie or those who view pets as valued objects may be quite willing to stick with elaborate conditioning procedures, whereas the worried, dissatisfied, or minimally involved owner is much less likely to persist in solving a problem and may be seeking your advice only with the hope that some drug will cure the problem. You may find that it is not worth investing your own time in working up a case if the client is unlikely to follow your instructions.

Clients who own pets primarily for companionship may be so involved with their pets that it causes some problems. For example, they may not be forceful enough to assume a dominant position over the dog. Instead, such a client may acquire a cat to keep the dog company and wonder why the problem is not solved. Other problems may arise with the low-involvement type of owners who leave their dogs continuously isolated from social contact.

Clients sometimes attempt to handle behavior problems by being overly anthropomorphic. A common example is when the client believes the dog is acting "guilty" when its behavior is simply submissive. The submissive behavior is the dog's innately determined way of stopping or attenuating aggression or punishment from the dominant individual, namely the owner. Often a dog owner reports that the dog knows it was doing something wrong because it acts guilty as soon as the owner comes in the door. In fact, many people say that they know a misdeed has occurred before they see the evidence, because of their dog's behavior. The dog, however, is most likely reacting to the owner's searching and asking behavior; anticipating punishment, it acts submissive and we interpret this as an expression of feeling guilty.

In another sense, there is a definite place for anthropomorphism in understanding animal behavior. Some dogs and cats are subjected to the same adverse early experiential effects and later abuse as some humans. Examples of such abuse include the mistreatment of puppies, the early separation of a mother from her young, and the isolation of newborn littermates from each other—circumstances that occur because pets are under the control of people. Keeping in mind the disastrous behavioral effects of repeatedly isolating a child from ordinary human contact or abruptly removing a child from its mother and siblings to be raised in an orphanage, the anthropomorphic perspective of animal behavior is at times useful. Once we recognize that young animals also have sensitive developing nervous systems, we can understand why behavioral problems are common.

Studies of child psychology have established the value of being consistent in meting out punishment or rewards: rewarding a behavior at one time and punishing it at another may produce emotional disturbances. The same holds true for pets. To punish a behavior such as barking at strangers at one time while rewarding and encouraging it at other times, results in confusion and slow learning. Allowing a dog out of a kennel on weekends but keeping it confined during the week is another example of inconsistent treatment that may lead to problems.

When we deal with problem behavior in pets, we must, of course, work with the owners. Often the therapeutic recommendation is formulated with the owner in mind. Some clients may be able to use the direct approach of meeting the aggressive dominance challenges of a dog with force. It is of no use, however, to tell a quiet, timid person to mete out severe punishment to a surly dog, even if this approach is indicated. Rather, an indirect approach to dominance, such as controlling affection (see Chapter 3), may be appropriate.

With some clients who are overly anthropomorphic with their dogs, we may need to emphasize over and over again that a pet's behavioral problems cannot be approached in the same way as a human behavioral problem. Other clients may understand more about behavior so that such explanations are superfluous. With these clients we may need to spend more time on the details of the therapy, or risk the clients' doing too much on their own and perhaps spoiling a well-planned therapeutic regime.

It is the client who decides that a behavioral problem exists. The client also sometimes plays a role in eliciting and maintaining the problem. The degree of success in treating a problem usually depends on the consistent effort of the client. Indeed, the client is the most critical component of behavior therapy.

Human-Pet Interactions

The close bond between humans and their pets has increasingly become the focus of clinical and research attention, as we see programs emerge that provide contact with pets for nursing home residents or patients in children's hospitals, and horseback riding for physically and mentally handicapped children. Practitioners see frequent evidence that pets enrich people's lives. They also see that the presence of a pet may heighten a stressful situation within a family or disrupt human social relationships.

Human-Animal Bond. Pets occupy a central role in the family, and family members often grieve when pets become ill or die. The bereavement can even be extreme enough to require psychiatric treatment. However, just as families can experience sadness with their pets, they can also be strengthened by their relationships with pets.

As people get older and are more likely to live alone, a pet can be a valuable source of closeness and interaction, helping people to continue their contact with friends and relatives. Pets can also lessen the serious problems of loneliness and isolation.

Pets are like children; they serve as social lubricants, facilitating interaction among adults. The realization of the value of pets has led to the human-animal bond movement, which recommends incorporating pets into nursing homes and hospitals to enhance the mental health of patients (Corson, et al., 1977). Pets require care and interaction, and are undemanding, uncritical friends. Patients become less withdrawn, since pets make them feel needed.

The greatest value of pets is companionship and having something to care for. In fulfilling these functions pets preserve health, particularly among elderly people. They provide people with something to touch, to keep them busy, to direct attention, to provide exercise, and to increase safety on the streets (Katcher and Friedmann, 1980).

The human-animal bond becomes dramatically obvious in the case of a seeing-eye guide dog for the blind, when a person is dependent on an animal, both in time of danger and for ongoing support. Hearing-ear dogs can now notify deaf human companions of a door bell, a telephone ring, or a fire alarm. Work is also beginning with monkeys that could potentially provide assistance to quadriplegic people by feeding them, turning pages, and adjusting light switches and doors.

Disruptive Effects. The attachment between people and their pets sometimes involves a remarkable amount of accommodation by the owners (Voith, 1981). Pets can provide essential companionship for people, but they can also be disconcerting and disruptive to human relationships; a point being studied by Simon (1984). Even a person living alone may select a pet with behavioral characteristics that are incompatible with his needs.

A psychotherapist treating a person reporting problems involving pet behavior might address the person's role in the problem. A therapist can directly deal with a person's tendency to relate to his pet while excluding people from his life, or to direct the most significant emotions toward a pet rather than family members. These issues are in the province of psychotherapy, not clinical animal behavior, but practitioners will often see evidence of personal disruption and acrimony being increased by pets.

A number of cases we are familiar with illustrate this point. Jim, a doctoral student, loved his Siberian Husky wolf-dog, Trinka. Jim spent most of his time training Trinka, to the point of neglecting his studies. Jim sought behavioral therapy for his dog because he wanted Trinka to enjoy being with him more, to like being inside the house, to eagerly interrupt a run when called. For her breed (half-wolf), Trinka was unusually affectionate and well-disciplined, but Jim was hurt that she was not more demonstrative in her love for him.

Two other examples illustrate some effects pets have on couples. When Sam and Sue were married, Sue had a Cocker Spaniel. Two years ago, Sam acquired a German Shepherd, which has recently begun to attack the Cocker Spaniel. Sue is clearly irritated with Sam, because his dog is beating up on her dog. Sam is uneasy that Sue's dog is getting so many extra privileges and that his dog is to blame. Divisiveness in Sue and Sam's relationship is increased by the aggression between their pets.

Sarah and her husband have four cats who live indoors. For the past two years there has been a daily spraying problem in the house, resulting in ruined carpets, kitchen cabinetry, and furniture, much of which has already been replaced once. Castration and progestin treatment have not helped solve the problem. To her husband's chagrin, Sarah wants the cats to continue sleeping in the master bedroom and would not consider limiting the cat's access to the whole house.

In each of these cases, the feelings and wishes of both of the owners were not considered in the selection or management of their pets. The pet, or its sounds or smells, may be extremely aversive to one family member but totally acceptable to another.

By the time the problem is presented to the practitioner, the pets, however inappropriate for the needs of the entire family, have already been selected. The practitioner's challenging task is to work with his clients on the management of the pet to create a more agreeable solution for the entire family.

Behavioral Therapy Management

The time needed to treat problem behavior in pets is much different than human behavioral problems or disorders. It is not uncommon for a human problem to require weekly sessions of 50 to 60 minutes for the duration of a year or more before progress is evident. When treating an animal behavioral problem we must produce results in one or two visits. In general, animal behavioral therapy carries a higher success rate than human behavioral therapy, since an animal's problems are often more clear-cut and can sometimes be managed in a 15-minute office visit. This is true of spraying in cats and problems in dogs that can be treated with progestins. This short office visit can often be sufficient to make a tentative diagnosis of attention-getting behavior and to give a client instructions on how to obtain additional information to confirm the diagnosis. In these cases, this short office visit can be followed by a longer session in which 30 minutes can be spent with the client. The longer visits might be scheduled just before lunch or at the end of the day to allow for time flexibility.

A particularly advantageous aspect of animal behavioral therapy is that few situations require immediate attention. Thus, a busy practitioner can schedule appointments for slower days or times when extra hospital help is around. During an appointment at the end of a slow day, you and the client may sit down for an hour and discuss a problem. The income for this time is not as great as that for surgery or general medicine; on the other hand, it is time spent at your discretion and convenience, and the therapy, whether successful or just an obviously honest effort on your part, is an effective public relations procedure.

To save time in taking the pet history, the office receptionist, an animal health technician, or some other hospital assistant can sit in on the longer visits and record the history. The assistant can also take the responsibility of calling the client to ensure compliance with recommendations and to answer questions.

Most problems lend themselves to a schedule that includes an initial office visit of 30 minutes followed by two progress-check visits of 15 to 30 minutes at one- to two-week intervals. In Chapter 2 we discuss the aspects of the medical interview that are useful in behavioral therapy.

PURPOSE OF THIS BOOK

This book is written for the small animal veterinarian who wishes to acquire some expertise in treating behavioral problems, and for the individual without the veterinary degree who has academic training in animal behavior and animal biology and plans to practice behavioral therapy through a veterinary liaison.

The role of the veterinarian is crucial in behavioral therapy because treatment of behavioral problems is multidisciplinary. Approaches require an integration of a physiological and medical background with the psychological and animal behavior fields. Veterinarians are able to diagnose disease processes in which behavioral changes may be secondary, and they are the most familiar with the physiological and medical aspects of treatment.

The danger in the emerging field of behavioral therapy is that some individuals who lack medical training may use an approach based only on one discipline, such as behavioral conditioning. This book incorporates knowledge from several disciplines that contribute to behavioral therapy.

SCOPE OF THIS BOOK

Treating a behavioral problem requires an understanding of the areas that contribute to the disciplines of behavioral therapy. These areas include the biological and naturalistic area, the physiological and medical areas, and the psychological and behavioristic area.

Biological and Naturalistic Area

Biologists and naturalists have traditionally sought to understand genetically acquired aspects of behavior from the standpoint of the adaptive value for the animal in the wild. Since the behavior of dogs and cats includes many behavioral patterns that were vital for survival in their wild ancestors, this field is particularly important in understanding behavioral problems. Some innate behavioral predispositions may cause problems when they occur in a domestic setting, such as various types of urine marking, hunting behavior, and aggressive behavior. In the domestic environment, these behavior patterns are no longer adaptive, and in fact, may be undesirable.

Other innate behavioral patterns are highly desirable and even vital to successful pet care. Dominant-subordinate relationships are important in establishing dog obedience. It is necessary to understand this aspect of natural canine social behavior so that owners can reinforce rather than counteract their pets' innate behavior. Were it not for an innate tendency to keep their nests or dens clean of eliminative products, dogs and cats would not make satisfactory pets. Problems with eliminative behavior require using an animal's natural "wired-in" behavior, and not basing training on conditioning alone.

We must also be aware that, in some animals, behavioral patterns may have been "weakened" through our domestic breeding practices, which remove natural selection pressures. Unfastidious behavior no longer necessarily leads to parasite infections, because we treat cats and dogs for intestinal parasites. Natural selection is also overridden when we provide artificial maternal care to young that are neglected by a mother. The young of "poor" mothers are then allowed to survive, and continue to breed a line of poor mothers. No amount of experience, teaching, punishment, or behavior modifications will transform such animals into good mothers.

Understanding the innate aspects of a certain behavior may dictate a certain approach in therapy. We can take advantage of an animal's innate predisposition to help dogs maintain a strong dominance-subordination relationship rather than fighting, or to eliminate a dog's house soiling behavior by relying on its innate tendency to keep its "den" clean.

We can also treat problems by altering the neural substrates for the behavior. By castrating male cats and dogs, we change undesirable male behavior by eliminating testosterone influences upon the brain. The recent introduction of olfactory tractotomy to treat persistent urine spraying in cats relies upon altering the olfactory input to the brain that evokes urine spraying.

Physiological and Medical Area

Problem behavior may be secondary to a disease or abnormality in one or more organ systems. Brain disorders are an obvious case in point. Some forms of psychosomatic epilepsy involve changes in emotional behavior. Traumatic injury or irritative processes in the brain can produce abnormal behavior along with other clinical signs. Problems with vision have been known to produce otherwise unexplainable occurrences in aggressive behavior. Diseases that are not obvious, such as arthritis or prostatic inflammation, but cause pain, result in irritable aggressive behavior. Abnormalities in the endocrine system may also lead to behavioral disorders. These examples illustrate the importance of a differential diagnosis in examining the etiology of a problem behavior and the value of a clinical work-up to diagnose the cause.

Some problem behaviors that look as if they have a pathophysiological basis are actually learned behavioral patterns that are attention-getting. Coughing, lameness, "hallucinations," self-mutilation, and fits have been detected as attention-getting behaviors, but the total work-up of the problem requires a combination of negative clinical tests and an understanding of the behavioral interaction between the owner and the pet.

In addition to contributing to the diagnosis of some behavioral problems, a physiological-medical approach is involved in several types of therapy. The behavioral indications of castration, progestin treatment, and neurosurgery fall into this realm. Psychoactive drugs, especially tranquilizers, may be used as an adjuvant to behavioral conditioning procedures such as desensitization. The choice of drugs and the monitoring of adverse side effects require attention from the medical field.

Psychological and Behavioristic Area

Historically, this area has reflected the efforts of those interested in understanding animal learning. Some useful principles of learning have been derived from work on laboratory animals, and often these are applicable to the diagnosis and treatment of behavioral problems. We are aware of the learning involved when we teach dogs or cats tricks, but we sometimes fail to realize that our pets are learning all the time. Animals may learn undesirable behavior because there has been some payoff, either in terms of the owner's affection or more tangible rewards such as food or water. An understanding of the principles of learning and behavior modification is useful not only in devising techniques to teach cats certain tricks, but also in eliminating undesirable behavior. The principles of counterconditioning, extinction, desensitization, aversion conditioning, and even the appropriate use of punishment, all of which are explained in Chapter 22, have been derived from basic work on animal learning.

Keep in mind that there are limitations to the use of behavior modification techniques. Urine spraying in cats or fighting between male dogs may be more easily altered by physiological-medical approaches than by conditioning techniques. Some behavior modification techniques may also be facilitated with tranquilizer therapy.

The techniques used by professional dog trainers include reward and punishment, as emphasized in the behavioristic approach to problem behavior, but trainers also rely on the innate behavioral reactions of dogs to accept a subordinate role when they use various techniques to gain dominance over them. Dog training is really an art based upon experience and highly individualized techniques.

ORGANIZATION OF THIS BOOK

This book is organized into three sections. The first section lays the groundwork for beginning behavioral therapy, including conducting the interview. The second section deals with treating specific problems in dogs and cats. The third section covers general concepts or principles, such as the rationale for the use of conditioning procedures, hormonal alteration, and drug therapy, and the selection and rearing of pet dogs and cats.

We anticipate that the chapters of the second section will serve as reference sources to solve specific problems, and that the chapters of the third section will provide useful background information. Practitioners new to the area of behavioral therapy may wish to schedule a short initial visit or ask the client to describe the problem over the telephone. When seeing the client a few days later, you will then have had a chance to read the appropriate chapter.

Chapter 2

Medical Interview and Case History Assessment in Behavioral Therapy

In general medicine and surgery, the medical interview and case history assessment are important in developing diagnostic approaches and therapeutic possibilities. These aspects of the case work-up are even more important when dealing with behavioral problems, because in many instances they provide the only source of information for the diagnosis of a problem and the recommendation for therapy.

During the first office visit the client describes the behavior and often attempts to give an evaluation of the problem. As the practitioner, you also attempt to diagnose the problem and evaluate the cause. Part of this evaluation might involve the question of whether the behavioral problem is secondary to a disease or abnormality in one of the organ systems, such as the brain, visual system, or endocrine system. The initial office visit may result in a decision to observe the behavior by hospitalizing the animal for a day or two. More often, the animal is not hospitalized and the client is given some instructions, and a progress-check appointment is made for a week or two later. Thus there is time, while the animal is hospitalized or between the initial and follow-up office visits, to consult this book and other sources regarding treatment for specific problems.

In this chapter we suggest how to get the most out of an office visit. Suggestions are made for evaluating a problem and forming a tentative diagnosis. Also, general guidelines for implementing therapeutic approaches are given.

GETTING THE MOST FROM AN INTERVIEW

The Client's Behavior

Our main contact with clients is through the office interview. An interview is a conversation with a purpose. When interviews lose this orientation, the practitioner's time is wasted and the client may feel cheated or dissatisfied. Despite this purposeful orientation, the practitioner must also be prepared to accommodate emotional situations. Most clients are distressed when they see us, and we must be sensitive to this distress. Although mismanagement, lack of skill, or ignorance on the part of the client may have caused the behavioral problem, it is important that we not convey a moral judgment about the client's behavior. Making judgments is a sure way to cause clients to withdraw. Most clients respond to compliments such as, "Well, you certainly are taking care of your dog," or, "It was smart of you to come in when you did."

The initial interview should be arranged to provide as much information as possible. It is usually helpful for a client to bring in the entire family, or at least the family member who is particularly attached to the pet.

The pet should also be brought into the office. Occasionally clients ask if the animal should be brought in, apparently believing that they will simply be discussing the problem. By watching the pet interact with the client, you gain first-hand information about how dominant the owner is over the pet. Alternatively, you may see that a dog is really a lap dog despite the owner's statement to the contrary. These observations may help in diagnosing some problems.

By having more than one member of the family present, you may learn about differences within the family in attitudes toward the pet. Does one family member want to get rid of the pet and another want to do everything possible to keep it? This usually means that the resolution of the problem will be more difficult, since all family members are not pulling together. Even when all family members are in agreement about solving a problem, having several of them present allows you to explain the instructions to all of them.

If the problem involves an interaction between two dogs, such as fighting, both dogs, not just the "culprit," should be brought in. This allows you to watch the interaction between the owners and the dogs to determine the degree to which the owners may be causing the problem.

The Practitioner's Behavior

The interview is the most important factor in evaluating and resolving problem behavior, but all interviews must be conducted under the constraints of a busy schedule. It is important to be aware of interviewing skills and techniques that play a role in efficiently conducting a medical interview. The practitioner who seems preoccupied or distant will not be maximally effective at obtaining the information he needs and will make a poor impression on the client.

Some forms of body language can give clients an impression about your feelings, just as you might use the same signs to read the client. For example, arms and legs tightly folded suggest a closed, withdrawn attitude. Leaning back with your hands on your head suggests a supercilious or superior attitude, which can prevent the other person from opening up. Leaning slightly forward with arms and legs relaxed is the most advantageous posture to invite others to speak. Eye contact is also important. We all realize this, but when we have both animals and people in the examination room, it is easy to constantly look at the pet while talking to the client.

You can physically arrange the examination room to facilitate your eye contact with the client. Putting the examination table to the side of the room, rather than in the middle where it stands between you and the client helps. Placing chairs for the client next to your chair and desk, so that you are almost forced to look at the client, may also be helpful. Placing stuffed toys on a shelf in the examining room can provide a diversion for young children so that the children do not disrupt your visual contact with the client.

It is best to occasionally paraphrase what the client has told you about the pet. In this way you allow the client to check your understanding of the problem, and the client knows that you are listening.

When talking to a client, the use of occasional compliments such as, "Wasn't that nice?" generally facilitates the interview and future relations with the client. If a client was neglectful or in error in the treatment of a pet, do not criticize. Even a question such as, "Why did you do that?" implies criticism.

Most successful practitioners treat a client and the client's pet as though they were very special, although they may have seen hundreds of such patients. The unruly dog that misbehaves because a client is not forceful enough may be a frequent and irritating behavioral problem, but you will get further by giving that client and dog particular attention than by emphasizing how common the problem is.

If you are upset about something that has nothing to do with the client, it is often better to tell him. Feelings are hard to mask, and tactfully admitting your feelings allows the client to understand the source of your irritation, rather than assuming that he or she is the source.

The style of your own questions has a role in the success of an interview. Open as opposed to closed questions allow the client some latitude, and may reveal information you might not have considered. For example, to ask "When does this occur?" rather than "Did this occur yesterday?" leaves room for the client to give you more information than you might otherwise obtain. The question "Well, how are things going?" is more open-ended than "Are things improving?"

The use of indirect instead of direct questions in sensitive areas is also wise. If you would like to recommend castration of a pet, for example, but feel the client may be sensitive to this operation, an indirect approach such as, "I was wondering about castration," leaves you more room to ease the client's emotional response than the direct approach of "How about castrating Peppy?"

STAGES OF THE MEDICAL INTERVIEW

Whether the office visit lasts 15 minutes or 60, it is best to approach the interview in stages. The stages are (1) the opening, (2) the development and exploration, and (3) the closing. This orientation focuses your attention on the necessity of efficiently progressing through the information-gathering process so that in your closing you have adequate time to give instructions to the client.

Before beginning an interview, some brief preparation may be necessary. The physical setting of the furniture should be such as to encourage eye contact with the client rather than centering attention upon the pet. Before going into the examination room, collect yourself, take a deep breath, relax your mind, and be prepared to give full attention to the client.

If you have set aside 30 to 60 minutes at the end of a day for a consultation concerning a behavioral problem, you might set the stage for a more relaxed session by offering the client a cup of coffee and having one yourself.

The Opening

In addition to exchanging pleasantries and greeting the dog or cat, try to evaluate the type of attachment the client has to the pet. Is it a strong emotional attachment, or one in which the client's interest is primarily economic? The client may want to breed the dog, but not if the dog's problem behavior seems incurable or gentically linked. Some clients are on the verge of getting rid of their pets and are more or less giving the pet one more chance—perhaps for the sake of some other member of the family. They may, in fact, be hoping that you will tell them the problem is insoluble; in essence, they may want your permission to have the pet euthanatized.

What is the Client's Goal? Assuming the problem appears to be solvable and the client will conscientiously follow any suggestions you make, your next task is to define the problem as precisely as you can. Determine what the client's goals are. You may find it useful to ask, "If I could give your dog a drug and produce a change, how would you like your dog to act?" For a dog that is aggressive toward children, the most desirable outcome may be for the dog to love children. While this goal may not be attainable, thinking about it and emphasizing it helps to focus your attention on the types of information needed and the types of behavior modification indicated.

Prioritize Problems. Sometimes owners will complain about more than one behavioral problem. In these instances it is usually best to focus on one problem at a time, unless all the complaints seem to be related to a single causal factor. Explain to the client that improvement in one problem area is often accompanied by improvement in other problem areas as well.

Development and Exploration

This portion of the interview provides the material for your diagnosis and points the way to treatment options. Several issues must be considered, depending on the nature of the problem and the pet's household.

The Problem Context. Once the specific behavioral problem is identified, it will be necessary to determine whether any specific stimuli or situations, such as the time of day or location, seem to be related to the occurrence of the behavior. Attempt to determine if any rewarding contingencies are maintaining the behavior. Does the behavior seem to be primarily learned, or is it a reflection of the animal's innate behavioral tendencies? For example, excessive barking may occur because the dog gets a frequent payoff, such as attention or being let into the house, but it may also be due to the natural alarm tendencies of dogs in response to noises and movements outside the home.

When is the Pet Good? Just as important as identifying when an animal is misbehaving is finding out when the animal is good. This information gives you an idea of what conditions might be explored to help solve the problem.

The Rambling History. Since solving a pet's behavioral problem usually involves the client's changing his or her interactions with the pet, it is necessary to learn not only about the pet's symptoms, but also about how those symptoms relate to aspects of the client's behavior or way of handling the pet. Thus, the interview for a behavioral problem usually involves more verbal communication with the client than a typical medical interview. Also, since the client usually does not know which variables relate to the problem, the critical pieces of information are often not presented in a simple organized sequence. The information may emerge from a long stream of anecdotes. The practitioner must be willing to listen and prompt the client with questions until the full picture seems to have emerged.

Early Experience and Genetic History. During the initial interview, most of the case history dealing with early experience will be obtained. To save time, it may be advisable to approach the early history of the pet only when it is indicated by the type of behavioral problem.

A genetic predisposition may exist toward certain behavioral characteristics. Questions regarding behavior of the dam, sire, and siblings may clarify the degree to which a genetic predisposition is present. It is important to consider the behavior of several related animals rather than just the dam, and it may be necessary for the client to contact owners of the sire and littermates later. Flank-sucking in Doberman Pinschers, narcolepsy in Poodles and Doberman Pinschers, and wool chewing in Siamese cats are abnormal behavioral problems that evidently reflect a genetic predisposition, but for which no organic disorder has been identified. Bear in mind that littermates share not only some common genetic background, but also similar early experiences.

Indications of early mistreatment may be significant, especially excessive punishment or teasing when the animal was a kitten or puppy. Some types of behavior, such as excessive timidity or aggression, may also be related to a young animal's being practically ignored by people. Avoid expressing an accusing tone while asking the owner questions about adverse early experiences, since the answers might become more defensive than informative.

Information regarding the source of a pet is also useful. Dogs and cats obtained from animal shelters, pet shops, or large breeding operations may have been isolated or ignored and are more likely to develop behavioral problems than animals obtained from a family breeder where kind treatment and plenty of affection was the rule.

Client-Pet Interactions. The behavioral interactions between people and their pets play an important role in the initiation and maintenance of a number of behavioral problems. People may condition undesirable behavior by rewarding an animal inadvertently. Difficulties with aggressive behavior in dogs may stem from the owner's lack of adequate dominance over the animal.

Differential Diagnosis. In the initial evaluation, you must rule out the possibility that the behavior reflects an underlying brain pathology or an abnormality in other organ systems. Sexual impotence in a breeding male, disturbances in feeding and drinking behavior, and irregular, unpredictable and unprovoked bouts of aggressiveness are examples of abnormal behavior that may be related to abnormal conditions in other organ systems. Clients often seem to expect that a behavioral problem is due to a brain tumor or other brain abnormality, but this is uncommon. The most common examples of abnormal behavior related to a known brain dysfunction are the changes in emotionality and aggressiveness associated with psychomotor epilepsy.

An extremely painful condition may cause an animal to threaten or bite if the tender area is handled. Acute inflammation of the prostate gland is an example of a painful disorder that may not be immediately obvious. A dog may tolerate handling by a dominant owner, but it may act irritable or aggressive if handled by others. Sometimes aggressiveness initially induced by a painful inflammation persists even after the inflammation has been completely resolved.

Some behavioral problems are normal innate responses reflecting a "mismatch" of the animal's natural behavioral predispositions to living in a confined household. Examples are urine spraying in the house by cats, urine marking in the house by dogs, furniture scratching by cats, objectionable sexual behavior, fighting between males, and roaming. Examples of problems caused inadvertently by clients are dominance by a dog over members of a family, fear biting, household destructiveness, and attention-getting behavior. Through our interaction with pets in early life, we may cause some dogs to be overly socialized to people, which makes them act excessively timid toward other dogs.

Directly Observing Behavior. Nothing can substitute for actually seeing a dog or cat's behavioral problem to understand what the client is talking about or to examine the external cues that may activate the behavior. One can observe behavior initially when the dog or cat is brought in for the interview and secondly during hospitalization. During the office visit the behavioral interactions between the animal and members of the family can be observed. This is useful, for example, when a dog is friendly or strongly attached to one member of the family but somewhat aggressive to another member.

When two or more animals are involved in a problem such as fighting between dogs, you can evaluate how the client interacts with each of the two animals if both dogs are brought in for the initial visit. During the office visit it is usually possible to assess the degree of control, or dominance, that an owner has over the dog.

It may be advisable to hospitalize the animal to observe undesirable behavior first-hand, especially when the animal is reported to display a highly unusual behavioral pattern. Hospitalization is often useful to determine whether an unusual behavior has a physical basis or the owner "rewards" the behavior by giving the animal attention or affection. The absence of the behavior when the animal is in a cage by itself (and unaware that anyone is watching) is an indication that the behavior is maintained because of its attention-getting abilities. If the hospital represents a pronounced change in environment, however, the pet may not show the problem behavior even if there is a physical cause for it. Under these circumstances it may be necessary to visit the owner's home to observe the behavior.

Closing Statements

The closing is, without question, the most difficult part of the interview. You must present your diagnosis or assessment of the problem, an overview of treatment possibilities, and specific instructions until the next office visit.

Do not rush through this stage of the interview. It is better to cut off or save some of your closing comments for the next visit than to be caught short of time and risk being misunderstood.

Choosing the Right Treatment. Interactions between a client and his pet, as observed during the interview, may affect your choice of treatment. For example, it could be clear that the most expedient solution to a problem of dominance by a dog is punishment of its aggressive acts. However, some dog owners are unable, either emotionally or physically, to administer enough punishment to a dog to obtain sufficient dominance and control. Such clients might be advised to enter into an obedience class where they can be assisted by an instructor. On the other hand, it may be necessary to forget about the use of physical force altogether, and help the client gain control over the animal by asking him to restrict his attention to the animal at all times except when he makes a demand of the animal and the animal performs.

This is the time to consider the relationship of the pet with various family members. Is it disliked by some and loved by others? Is the dog draining the affection of the husband away from his wife? Are the pet and its problem keeping family members from addressing important difficulties in their family relationship? Sometimes therapeutic suggestions can go in the direction of helping to reduce divisiveness in a family. This happened in one instance in which the presenting complaint of anorexia could not be treated until we had solved the problem of mounting by a dog of the wife and daughter. At times your advice can inadvertently add to the divisiveness, as when taking sides with a husband or wife in solving a problem about which the couple have a disagreement.

Keep in mind that you may see possibilities for the client to avoid or reduce a problem by management rather than specific behavioral therapy. One might suggest confining the pet to a certain location for certain misbehaviors. Problems of urine spraying or inappropriate elimination may be lessened if the pet is outdoors when the owners are not home or are asleep. The potential for destructive behavior also depends on where the pet is left when unobserved. If the solution to a behavioral problem is based on avoiding the problem, the client must be willing to change his management procedures. If the client, for example, responds that the pet is too old to go outside, or is accustomed to sleeping in the bedroom, other alternatives must be sought. Accurately assessing the client's willingness to modify routines with a pet is one of the most critical tasks.

Pharmaceutical Treatment and Hormonal Alteration. This book reflects an approach that emphasizes behavioral modification rather than pharmaceutical treatment. Generally, behavioral approaches are powerful tools and lack the side-effects and experimentation with dosage level adjustments that are necessary with tranquilizers. There are, nonetheless, definite indications for the use not only of progestins but also tranquilizers. These are discussed in Chapters 23 and 24, although a couple of examples are worth mentioning here.

Occasionally an animal is so anxious that it is impossible for the owner to conduct the training sessions needed in a desensitization program. Tranquilization can block the emotional response long enough to begin systematic desensitization. At times, an animal facing a specific occasion that will be terrifying, such as Fourth-of-July fireworks, or a long cross-country trip, may benefit from tranquilization, when it is not appropriate to carry through a desensitization program. For these occasions it is wise to rehearse and test the dosage level before the critical event. A tranquilizer is especially indicated for cats that engage in extreme withdrawal behavior.

Hormonal modification by castration or administration of a synthetic progestion are effective ways of modifying behaviors that are sexually dimorphic, such as spraying, roaming, and fighting in cats, and urine marking, roaming, fighting, and mounting in dogs.

Delaying the Decision. There are times when we find it difficult to devise a therapeutic approach. To buy some time for consulting references or other colleagues, the practitioner can ask the client to collect more information, such as the occurrence of the behavior in the dam or siblings, or the occurrence of the behavior in the client's absence, and to return to the office in a week for a discussion of treatment.

The Client's Rehearsal. In most behavior cases, the challenge is to lead the client toward a new pattern of interaction with his pet. This type of teaching, like others, is best learned with practice. Why not encourage the client to practice the new procedure in your examination room or office while you are there to offer advice? The client may feel silly rehearsing a conditioning approach in your office, but it will enable him to feel more confident when he actually begins the training session at home. A few examples are worth mentioning here. If a family with two dogs is to transfer the role of the favored and privileged pet to the other dog, the shift will come more easily after the family members have actually practiced interacting with the dogs in your office. If you are recommending interactive punishment for aggression directed toward the owner, why not evoke the aggression and help the client punish the dog in your examination room? A rehearsal with the client gives both the client and you an opportunity to see whether you share the same idea about the details of the training procedure. This is a good time to work out the specific variables for the initial training sessions, such as duration, distance, location, and participants.

Euthanasia as a Solution. This option may surface in the minds of clients after they have spent some time trying to resolve a problem or after hearing of the time commitment needed to resolve a problem. The most frequent nonmedical reason for requesting euthanasia for a pet is for behavioral problems. But keep in mind that it is more emotionally traumatic for pet owners to accept euthanasia of their pets when it is performed because of some behavioral problem than when it is performed because of some type of terminal illness. Owners have a strong tendency to feel guilty or inadequate for possibly having caused the behavior. It may be necessary to emphasize the parallel between some types of behavioral problems and terminal or incurable disease, emphasizing that the client is not really at fault. Delaying a euthanasia decision by offering the client an alternative, even one that has a low probability of succeeding, is sometimes worthwhile. After making a further effort to correct the pet's behavior, the client knows he or she has tried everything and feels less conflict in making the choice for euthanasia. The performance of euthanasia brings up the prospect of dealing with the grief reactions of clients. This is dealt with in Chapter 1.

Client Noncompliance and Forgetfulness. Sometimes clients are reluctant to follow suggestions, especially when we prescribe affection withdrawal or interactive punishment. In such cases, suggest that they experiment with your suggestion for one week and come back and talk it over again.

We must all deal with the client's tendency to forget most of what we say. Studies on how much human patients tend to forget of the information given to them by physicians are shocking. Of course, the more patients are told, the more they forget. These findings on forgetting probably apply to our clients as well. Since we rely almost completely on a client's following our instructions, this is an extremely important concern. In behavioral problems, excluding those treated pharmaceutically or surgically, most of the improvement is instigated, directed, and main-

CASE 2–1

History. Shabby is a Toy Poodle who urinates on her owners whenever they pick her up. She does not misbehave when the owners are gone.

Diagnosis. Submissive Urination

General Evaluation. Shabby urinates when she is anxious. As treatment, she will be placed in anxiety-provoking situations only during training sessions when any misbehavior will be remotely punished. At other times avoid triggering her anxiety if at all possible.

SPECIFIC INSTRUCTIONS

1. Reduce Shabby's anxiety except during training sessions. Do not reach down toward Shabby or pick her up, or do anything that is likely to make her urinate except when another family member is present to give remote punishment with a water sprayer.
2. Allow Shabby to approach and contact family members who are sitting or lying still as long as this does not cause her to urinate. This is the main type of human contact she should get for the next week.
3. Schedule remote punishment sessions. One member of the family should pick up Shabby in a manner that will make her urinate. Immediately after the urination, another family member should squirt her in the back of the head with a water sprayer. Shabby is not to know who squirted her or where the water came from. The person picking up Shabby should act mildly surprised about the water and drop Shabby. The dog should be ignored for one hour.
4. Several water spraying sessions can be conducted each day. Rotate among family members as to who picks Shabby up and who does the squirting.
5. If Shabby does not urinate when picked up, continue to hold her, praise her, and give her a good treat, but do not overly excite her.
6. Progress check in one week.

CASE 2–2

History. This Miniature Schnauzer spends most of his day snapping in the air. He follows the owner from room to room during the day, but seems to focus his full attention on chasing and snapping at what one might consider to be imaginary flies.

Diagnosis. Attention-Getting Behavior

General Evaluation. Our conclusion is that the "fly-snapping" behavior is related to attempts to gain attention. The instructions below are designed to extinguish fly-snapping behavior and promote "good" behavior through positive reward.

SPECIFIC INSTRUCTIONS

1. Avoid rewarding snapping. When Tommy snaps, leave the room, or the immediate area, and completely ignore him.
2. When there is no snapping, reward Tommy. Give him food treats, praise, and affection when there is no fly-snapping for at least five minutes. This time period will be extended as he gets better. A water sprayer can be used to break the cycle of fly-snapping and to produce good behavior as long as he doesn't see where the water is coming from.

tained by the client. By changing one's own behavior with the pet, the client changes the pet's behavior. Since the client is a critical link to correcting the problem, it is essential that your communication with the client be absolutely clear.

Here are three concrete suggestions for dealing with forgetting. The first is to give the client a brief written outline of your instructions. Cases 2–1 and 2–2 represent two examples of written instructions given to clients. One set of instructions was for Shabby who urinated on the owners every time she was picked up. The instructions intended for Tommy deal with the complaint that he was snapping at imaginary flies. The diagnosis of attention-getting behavior was based on the discovery that Tommy never displayed the behavior when he felt he was alone; the instructions are those from the second office visit. These instructions were, of course, greatly expanded upon in our discussions with the client. You might use a hospital letterhead for the instructions with a carbon copy for your files. Instructions should be as specific as possible. It is too much to expect a client to work from a general theory to a particular situation; that is your job. The more simple and specific the instructions are, the less the client will forget and the greater will be the compliance with instructions.

The second suggestion is to schedule a series of three or four office visits at intervals of a week or two. This not only allows you to reduce the number of instructions required per visit, but gives you the opportunity to review and make adjustments in the instructions. After the first visit you may request only that the client obtain more information about when a problem occurs. You then have the opportunity to consult references about the problem before giving detailed instructions next time. It is also wise to call the client yourself or to have an assistant call to see how things are progressing between office visits.

The third suggestion, as indicated previously, is to have the client practice the instructions with the pet before leaving the examination room. If the client has rehearsed the procedure, he or she will be far less likely to forget the details of it.

Section II

CAUSES AND THERAPEUTIC APPROACHES FOR COMMON BEHAVIORAL PROBLEMS

In this section we deal with the most common behavioral problems of dogs and cats, which include aggression, fear reactions, activity problems, eliminative behavior, roaming and escaping, problems with feeding, sexual behavior, and maternal behavior. The therapeutic approaches for many of these problems require familiarity with the background information presented in Section I. This section is meant to be a quick reference source for guidance in treating the most common problem behaviors.

The chapter titles on cats somewhat parallel those on dogs, but it needs to be emphasized that the therapeutic approaches differ between the two species. The differences in treatment primarily reflect differences in the social behavior of dogs and cats. Dogs are drawn into social interactions with us because they have a social nature that goes back to the pack life of their wild ancestor, the wolf. Social animals tend to organize themselves in rank orders or dominance hierarchies, and once these are established, fighting is reduced. Conflicts that arise are resolved by a threat from one animal and then, usually, a submissive gesture from its opponent. Dog problems are often centered around the dominance hierarchy issue, and time after time, issues relating to dominance come up in problem behavior.

Cats, on the other hand, are basically asocial. They respond to us as sources of food and shelter or as mother figures, but not as substitute peers or pack members. Freely roaming asocial animals do not use the behavioral patterns of threat and submission to resolve conflicts easily or to reinforce social rank, and consequently, when animals interact, fighting is more frequent. In the wild, this system works for members of the cat family because cats tend to circumvent conflict by avoiding interactions with each other.

Since cats do not form dominance hierarchies and do not accept the domination of people in the same manner as dogs, it is useless to attempt to punish a cat as one would a dog. Cats do not show much in the way of submissive responses, but rather react to punishment, or any type of force, by escaping or fighting.

Frequently, cats are highly social and affectionate when relating to people. This is not a reflection of whether cats are innately social or asocial, but rather that they react toward people as a kitten would toward its mother, by kneading our laps and purring in our presence.

The fact that cat owners keep treating their cats as social animals comes up again and again. People get cats to keep other cats from being lonely and often end up with two cats living independent existences in the same household. Some people get a cat to keep a dog company, or a dog to keep a cat company. Whatever the life style of a cat, its underlying solitary nature surfaces from time to time, and problem behavior usually results.

A recent volume on the behavior of domestic animals (Hart, 1985a) presents a detailed coverage of the normal behavior of dogs and cats along with that of the other common domestic mammals.

When appropriate in the discussion of problem behavior, we have included examples in which a therapeutic approach has been used in a specific case. The instructions are those given to owners. Most of these samples are actual cases, but the names of the clients and patients have been changed.

Chapter 3

Aggressive Behavior in Dogs

The most frequent type of behavioral problem in dogs is aggressiveness. Aggressiveness, however, is not always an undesirable trait. One reason dogs were domesticated was because they protected their adopted human homes as vigorously as they would their native den. An aggressive watch or guard dog can be a real asset in a crime-ridden neighborhood. Furthermore, we admire a stately dog that is well-mannered around other dogs, but who turns and attacks an unruly dog that delivers repeated insults.

Problems arise because of the variability in aggressive behavior. We expect our dogs to aggressively keep out intruders, but not threaten or attack our friends. They should not growl at other dogs, but neither should they back down from a bully. A dog considered aggressive by one person may be just right for another.

THE NATURE OF AGGRESSIVE BEHAVIOR

In dealing with problems of aggressive behavior, it is necessary to tailor our therapeutic advice and recommendations to the client's wishes and lifestyle. The first approach is to determine the type of aggressive behavior. The aggressiveness displayed by a dog fighting with another dog over a bone is different from that displayed by a dog protecting its territory. These examples of competitive and territorial aggression are both different from the fear-related aggression that we see in fear biting, and the therapeutic approach to each of these types of aggression differs. The various types of aggressive behavior are discussed as an aid in understanding aggression and diagnosing problem behavior.

In nature, animals are most aggressive toward members of their own species. The classifications of aggressive behavior have been developed primarily to characterize intraspecific interactions. A classification adapted from Moyer (1968) is included here to familiarize the reader with aggressive behavior in wild canids.

A second classification, based on whether the aggression is toward other dogs or people, relates to the fact that dogs form dominance hierarchies and allow humans to enter their hierarchies. For treatment purposes, the critical feature of aggression is whether it is directed at dogs or humans.

In our discussion of the diagnosis and treatment of aggression, we will initially characterize aggression as human- or animal-directed, and then consider the type of aggression.

Types of Aggressive Behavior

The most widely accepted classification of types of aggressive behavior was developed by Moyer (1968) and includes the following: predatory, irritable or pain-induced, fear-induced, learned, intermale, and maternal. We have extended Moyer's classification, specifically for dogs, to comprise the following types: competitive, dominance-related, intermale, fear-induced, pain-induced, territorial, and idiopathic.

Competitive Aggression. Dogs may aggressively compete for a highly favored food, a bone, or some particularly desirable rest area. Fighting over attention from the owners is also common. The situation is simplified when one dog is clearly dominant over another, and conflicts are resolved by threats and submissive gestures and actual fighting is unnecessary. Owners should not expect dogs to share equally. When dogs frequently engage in serious fights concerning competitive issues, the owner should help establish or maintain the dominance of one dog over another. This type of aggressive activity generally ocurs in females as well as in males.

Dominance-related Aggression. In groups of social animals, peace is maintained by the existence of dominant-subordinate relationships. There is an inherent tendency for animals to strive for the top position. Thus, in a two-dog or three-dog household, a subordinate dog is ready when the opportunity exists to take the top position. This may involve a fight as the subordinate sees its opportunity to take over when the top dog becomes too old, gets sick, or loses the protective umbrella of the owner. Dominant-subordinate relationships are of value, since they prevent fighting when there is competition for food or the owner's attention.

Since dogs relate to people as they would canine members of a pack, there is the potential for dogs to attempt to assume a dominant position over their owners. Often the dog does this by threats or snapping when the owner tries to assert himself or herself. The dog may then become more aggressive to maintain its dominant position.

Intermale Aggression. Male dogs are innately drawn into fights with other males. Females do not show a comparable propensity for fighting with each other; hence, this behavior is sexually dimorphic. This is the only type of aggressive behavior that is clearly known to be altered by castration (Hopkins, Schubert, and Hart, 1976; Hart, 1947b; Hart, 1976).

Fear-related Aggression. When veterinarians and animal handlers are bitten, it usually reflects this type of aggression. Fear-related aggression is seen in females as well as males. It occurs in circumstances where an animal would escape if possible. Although aggressive threats and attacks are innate responses of a fearful animal being cornered, there is a learned element to this behavior. When people evoke a fear response in a dog, the growling, threats, or snapping are often reinforced, because when people are driven away, the dog's fear is reduced. The more we give into fear biting, the more an animal is affected by its pay-off. In nature, fear-related aggressive behavior will occur when an animal is cornered by an enemy, be it a conspecific or a predator. We do not see much of this on the domestic scene.

Pain-induced Aggression. When we administer pain to an animal, it has a natural tendency to bite back in self-defense. In fact, it is often pain that keeps a dogfight going. It is not recommended that people attempt to break up a dogfight by hitting the dogs, since the added pain may simply intensify the fight. We are all familiar with the dog that growls or snaps when it is touched on a painful inflammatory area. This aggressive behavior is protective and usually presents no serious problem. If the aggression persists after the inflammation is gone, a special desensitization program may be needed to eliminate the behavior.

Territorial Aggression. Dogs are famous for their territorial aggressive behavior. They seem to react to strange people as they would to strange dogs: as a threat to their territory. Territorial aggression can be a desirable behavioral pattern, but when dogs routinely attack friends, meter readers, or postmen, the behavior can be a serious problem. Most territorial animals are aggressive only to conspecifics entering their territory. In fact, the aggression of dogs toward human intruders is a rather unique interspecific example of territorial aggression.

Predatory Aggression. This classification refers to the natural tendency of carnivores to kill animals for food. We, of course, shape this behavior in our hunting animals by teaching dogs to retrieve, or to chase foxes or tree racoons. Problems can arise with the predatory behavior of some dogs when they attack chickens, cats, or sheep. Predatory aggression, of course, does not involve growling or snarling as a preliminary warning; hence this type of aggression differs in behavioral topography from other types in which threatening is common.

Maternal Aggression. Females of most species of wild animals are capable of intense aggression in defending their young and, in fact, will fight to their death. The physiological basis of this behavior is not understood, but it is believed to be partially a function of the hormonal state of the postparturient female. Because of our close association with dogs in most homes and kennels, this type of aggressive activity is only rarely a behavioral problem. The killing and consumption of newborn, or cannibalism, is a different type of aggression, and is dealt with as a problem in maternal behavior (see Chapter 11).

Learned Aggression. Animals can learn to become excellent fighters. We can also teach a dog to fight on command so that a full-blown vicious aggressive response can be brought under vocal control. Also, learning may be involved in shaping other types of aggression into more intense forms. In fear-induced aggression, for example, threats or snapping can drive away people that incite fear in a dog. If the people eliciting the fear decide not to allow the threats to chase them away, the dog may resort to snapping. If this works, the progression from threats to snapping represents learning.

Idiopathic Aggression. This is a term that can be applied to instances of aggressive behavior that appear to be truly abnormal and unexplainable by normal causes. Examples from the world of wild animals are nonexistent. The most striking case involves unprovoked vicious attacks on people in a dog's own family. In such episodes the dog is described as getting a glazed or distant look in its eyes, appearing not to recognize people in the family and following through with a vicious attack. These attacks may occur with little warning, and can result in injuries to the face, neck, and arms. The dog may be an affectionate and enjoyable pet at other times.

Factors Affecting Aggression

Most types of aggressive behavior are influenced by several factors, including early experience, genetic background, and prior conditioning or learning.

Socialization Factors. Dogs that have had little contact with people in early life, especially 3 to 12 weeks after birth, may be overly cautious and fearful of people (see Chapter 21). As adults, they may react to people with fear biting. Such dogs may get along fine with other dogs. On the other hand, dogs taken from their litters at 3 to 4 weeks of age, with little exposure to other dogs until adulthood, may be very aggressive toward other dogs. Their responses toward people can be quite acceptable. The tendency to fight with other dogs may reflect inadequate socialization in the use of threats and submissive postures, which allow dogs to settle conflicts without actually fighting.

Dogs that have not had a chance to become accustomed to young children early in life can be quite fearful of children later. This can be a problem in a childless family where a dog may have lived happily with a couple until the couple decided to have a baby. For people who have no children, and who wish to adopt a puppy, it is wise to have the puppy exposed to children on a regular basis.

Dogs often interact with people as they would another dog. Thus, the issue of who dominates whom invariably comes up. Probably the most important reason for taking dogs through obedience training is to make certain that the owners will be dominant over the dog; the tricks or procedures the dog learns are of secondary importance. The best time for obedience training is when the dog is a juvenile. Unruly dogs who start controlling their owners at an early age may create serious problems for the owners as adults. However, obedience training for an adult dog is better than none at all.

Genetic Factors. Some breeds of dog have a greater propensity to become aggressive than other breeds. Retrievers are known for their friendly temperament, while certain varieties of terriers are more easily induced to display aggression. In Chapter 21, we discuss breed differences with regard to some types of aggressive behavior.

Males versus Females. Most forms of aggression become a problem more frequently in males than in females. This is probably the result of the underlying differences in general aggressive temperament between males and females, stemming from differences in gonadal secretion prior to birth (see Chapter 23). It is questionable whether prepubertal castration would prevent males from becoming more aggressive than females. The sexually dimorphic differences are probably too well established even at the time of birth for prepubertal castration to be any more effective than postpubertal castration. Intermale aggressive behavior is activated at the time of puberty by testosterone secretion, and it is the only type of aggressive behavior that is predictably reduced by castration. Even so, castration eliminates intermale aggression in only about 50% of dogs (Hopkins, *et al.*, 1976).

Aggression Secondary to Other Disease Processes. Diseases of various organs may, on occasion, lead to the onset of aggressive behavior. Aggressive behavior resulting from brain lesions is rare, although it is possible for pressure exerted on the hypothalamus by a tumor of the pituitary gland to result in a gradual increase in irritability or aggression. Some visceral or metabolic effects will usually arise concomitantly with the aggressiveness. In some cases strange aggressive reactions emitted shortly before or after a seizure attack may be a symptom of temporal lobe or psychomotor epilepsy (Holliday, Cunningham, and Gutnick, 1970).

Serious visual impairment may predispose an animal to some types of irritability because of a dog's inability to recognize a person. An inflammatory process in joints, muscles, or internal organs such as the prostate gland may also cause a dog to act irritable and aggressive. The irritability may persist after the healing of the inflammation because of conditioning that occurred when handling was painful.

Use of Drugs to Control Aggression. Tranquilizers have been used to diminish aggressive responses while a dog is temporarily desensitized to the situations or stimuli that have evoked the aggression (see Chapter 24). Tranquilizers may also be used when dogs are being placed together to reduce fighting. When tranquilized, the animals can establish a dominant-subordinate relationship using milder forms of aggression, and the owners can work to help establish one dog in a dominant position. After a dominant-subordinate relationship is established, the drug dose is gradually lowered. Synthetic progestins are used to suppress intermale aggressive behavior when castration is not effective.

TREATING AGGRESSIVE BEHAVIOR DIRECTED TOWARD PEOPLE

Because we live so closely with our dogs, and enter their social structure (becoming a member of their pack as it were), we are subject to some of the kinds of aggression that dogs would display toward other canine pack members or toward an enemy or territorial intruder. Understanding some of the factors covered in the foregoing classification of aggression gives us some direction in dealing with specific human-directed aggression. For the sake of convenience in diagnosing problem aggression, we will initially discuss the typical kinds of aggressive behavior directed toward people, and then the typical aggressive behaviors directed toward other dogs. A dog will usually not display the same problem aggression toward both people and other dogs.

Dominance-Related Aggression

History. This form of aggression is especially common and relates directly to the behavior of the owners. One person's obedient and perfectly controlled pet may be another person's problem. Some dogs growl or snap at their owners, not out of fear or pain, but because they simply want their way. The dog may threaten or snap when attempts are made to put it outdoors, or move it from the bed. Sometimes the dog even appears to "dare" the owners by grabbing something like a shoe or glove and displaying the object. When the owner attempts to take the object, the dog growls and threatens.

In the hands of a person who assumes a more dominant role, such as a veterinarian or dog trainer, the dog may not display this aggressive dominance. The owners who suffer at the hands (teeth) of such dogs are often the type who tend to pamper or spoil their dogs.

Some probing into the history of dogs presenting this type of problem often reveals that the dog's behavior is fine 99% of the time. For the most part, the dog may accept the owner's role as boss. However, when the dog wishes to call the shots, it simply assumes its aggressive mode and takes the dominant role.

This type of problem is usually prevented when the owner takes a dog through obedience training early in life. However, having gone through obedience training does not assure that the dog will always be perfectly obedient.

Direct Therapeutic Approach. There are two rather different approaches to resolving this problem, one is direct and the other is indirect. The direct approach exploits the dog's natural tendency to assume a subordinate role once it learns that being dominant is not possible. In a pack, dominant-subordinate relationships are established and maintained by force or the threat of force. Physical force is a type of nonverbal communication a dog understands, and dogs certainly use force with each other.

Some dog owners do not understand the role of force in curbing aggressive dominance. Aggression directed toward the owner, family members, or friends is intolerable, and is one instance where severe interactive punishment is indicated. The direct approach of meeting aggressive dominance with severe punishment can only be handled by some owners. If in your judgment the owner is not physically or emotionally capable of dispensing the needed punishment, it is best to start with the indirect approach and work toward the direct approach. The goal must be that eventually the owner will be able to do practically anything to a dog physically without the fear of being challenged.

To apply the direct approach, the owner meets the dog's aggression with the appropriate force. Shaking by the scruff of the neck is sometimes appropriate for small dogs. One convenient technique for larger dogs is for the owner to punish aggression severely by using a choke chain. If the owner has tried control with the choke chain and the dog has learned to snap as soon as the owner reaches for the chain, some additional mechanical advantage may be necessary. If the owner ties a short rope onto the choke chain with a large knot on one end, it is then possible to manipulate the choke chain without coming too close to the dog's head.

Once the owner grips the choke chain, he must win the encounter. This may require being very severe with the choke chain and possibly hitting the dog as well. At this point some dogs may fight back, especially if they have already learned to be quite aggressive. The technique of stringing up dogs by holding them off the ground with the choke chain until the dog almost passes out has been recommended by some authorities as a procedure for the most extreme cases. If situations are staged where it is known the dog will act aggressively, then the owner can be prepared, with gloves and leather boots if necessary, to administer the punishment. In the sample case presented here, Pamela's owner staged mild aggressive encounters at meal times so she could be prepared to deal forcefully with the dog (Case 3–1).

Meeting a dog's force with equal force should not be considered cruel, when it is recognized that dogs use force with each other in the form of biting. The owner must be capable of winning the confrontation. There are, however, many dog owners who are physically or emotionally unable or unwilling to meet a dog's aggressive episodes with the kind of force that is necessary to win, or who feel they must treat their dogs the same way they treat their children. The indirect approach is recommended for such circumstances.

CASE 3–1

History. For the past year Pamela, a spayed female English Springer Spaniel, has snapped and bitten at people. She recently bit a houseguest and now regularly growls at the owner. Pamela has always been possessive of her food and other things.

Diagnosis. Dominance-Related Aggression

General Evaluation. Pamela is accustomed to intimidating the owner. In treatment, each instance of Pamela's threats and aggression will be punished.

SPECIFIC INSTRUCTIONS

1. Stage training sessions to elicit mild aggression each evening by presenting the food bowl and then taking it away. Punish Pamela vigorously if she growls. Leave a leash hanging from her choke collar to simplify safely correcting her.
2. When she is good, reward her by replacing the food bowl and giving her a favorite food treat.
3. Avoid provoking aggression outside of the training sessions, but if it occurs, punish her and put her outside. Do not take the subordinate role in any encounter if at all possible.
4. Work with the rest of the family in asserting dominance over Pamela by using the same procedures.

Indirect Therapeutic Approach. The indirect approach is useful for dogs that are so accustomed to fighting for dominance that they will never give in during a physical confrontation. Starting on the first day of behavioral therapy, the owners should completely withdraw all physical affection and normal feeding from the dog. Also, for the duration of the training period, they should avoid circumstances that are likely to evoke an aggressive response from the dog. If, for example, forcing the dog outdoors is likely to evoke a threat or a snap, then other ways must be found for getting the dog outdoors. Frequently throughout the day, and especially when the dog appears to want attention, the owner should call the dog over, issue a command such as sit or lie down, and when the dog responds favorably, give it lots of affection and a portion of its food ration. The dog is still able to get a day's worth of social contact and dog food, but on the owner's terms rather than its own. The dog is never rewarded unless it obeys the owner. Thus, the owner is in control and indirectly assumes a dominant position. If commands are given frequently throughout the day, the dog is constantly reminded of its subordinate position. The success of this approach depends on all members of the family stringently withholding affection from the dog except when commands are given. Food deprivation is a useful adjunct.

Once the procedure begins to have an effect on the dog, the owners should be advised to use a little force or "bullying." They can push the dog outdoors if it's a little slow in moving out, or jerk down on a choke chain if the dog does not lie down on command. Gradually the owners should increase the force with which they handle the dog until they are at a point where they can deliver considerable physical punishment without fear of reprisal.

In household situations when only one member of the family has trouble with the dog, all members of the family should ignore the dog except that person having the trouble, who should issue commands and be the only source of affection and food rewards.

Sometimes a dog with a particularly mild-mannered owner increases its aggressive reactions so gradually that the owner is slow to notice the problem. In one case (Case 3–2), a Manchester Terrier bit the owner several times before she requested assistance. Treatment included an initial two week period of affection withdrawal, when rewards were only given after commands had been obeyed.

Dogs may show an aggressiveness toward a boyfriend or girlfriend. In these circumstances, the dog's owner should completely ignore the dog and any petitions for attention for about two weeks. The boyfriend or girlfriend that was in the dog's disfavor then becomes the sole source of love and affection. The friend toward whom we wish the dog's attitude to improve is instructed to give the dog a command such as "sit" before petting or interacting with the dog. The dog, therefore, learns to respond in a subordinate way to this new individual.

CASE 3–2

History. The client does not strike you as someone who could be assertive with a misbehaving dog. Her 4-year-old female Manchester Terrier has bitten her several times when she has made the dog move from her bed, or when she has walked past the dog when it was sleeping on a pillow in the living room. Normally the client gets along well with the dog and is very fond of it.

Diagnosis. Dominance-Related Aggression

General Evaluation. The dog has developed an escalating pattern of aggression. Since it seems to difficult for the client to threaten the dog with force, we suggest treating the aggressiveness indirectly.

SPECIFIC INSTRUCTIONS

1. Starting tomorrow, and continuing for two weeks, completely ignore your dog unless you give it a command. She is not to sleep on your bed or be petted when she solicits affection. Withdraw food for 24 hours.
2. Throughout the day give the dog commands such as "sit" or "lie down." When she obeys give her plenty of praise and affection for about 30 seconds and give her a portion of her food ration.
3. Gradually increase your demands on her and start acting like a bully. Get more pushy with your demands.
4. Avoid circumstances that are likely to evoke aggression. "Losing" any confrontation will reverse the progress made during the affection withdrawal period.
5. Progress check visit in two weeks for next stage.

CASE 3–3

History. Mr. Taylor has no problem handling Matt, a 2-year-old Rottweiler, but Matt is disobedient with Mrs. Taylor. Matt jumps on furniture where he is not allowed, often carrying in his mouth objects that he should not have. Mrs. Taylor is afraid to correct him, even when he growls at her. He is particularly aggressive if she reaches for something in his mouth.

Diagnosis. Dominance-Related Aggression

General Evaluation. The goal is for Mrs. Taylor to control Matt. During treatment he will be punished for misbehavior and will be entirely dependent on her for affection.

SPECIFIC INSTRUCTIONS

1. Mr. Taylor should completely ignore Matt, giving him neither positive nor negative attention. Mrs. Taylor should also ignore him on the first half day. Then, through the second day, she should ignore Matt except when she gives him a command or wipes his mouth. Ignore Matt if he shows any aggressive behavior, but praise him if no aggressive behavior is shown. Repeat this frequently each day, and continue throughout the week. Withhold treats except when he behaves well. To aid in handling him, let him drag around a rope or leash tied to a choke chain.
2. During the week make increasing demands on Matt. Wipe his mouth more vigorously. Place a rawhide bone 2 feet away and then take it away, and then gradually decrease the distance.
3. Progress check in one week for the next stage of training.

The use of either the direct or indirect approach, or a combination of the two, as was used with Matt (Case 3–3), merits attention to detail.

Office visits should be scheduled at weekly intervals for the first three or four weeks for consultations. It is usually as difficult for owners to withdraw affection from their dogs as it is for them to dole out severe punishment when necessary. The office visits or telephone calls are necessary to remind your clients of the importance of sticking with the procedures. Later, a session with an obedience class is very useful in consolidating the reversal of dominance.

Competitive Aggression Toward Children: A Form of Sibling Rivalry

A well-mannered dog that has never given anyone a problem with dominance or aggression may develop a nasty temperament toward a child or new baby in the family. Suddenly the family members are no longer sure that they can trust the dog not to bite. Your advice may be sought after the dog has growled or snapped at a child. In this situation, the concept of sibling rivalry is relevant because the dog was the "only child" until a new baby arrived and started moving in on the dog's share of attention.

History. The dog has been a perfectly nice and lovable pet for the adults it lives with, and even gets along with the neighborhood children who are occasional visitors. Successful treatment of this problem allows these people to keep a lovable pet that might otherwise be subjected to euthanasia. The origin of this problem, and the solution for that matter, relate to the way in which affection and attention shape a dog's behavior. Often a couple may lavish affection on their dog, assuming the animal needs assurance that it has not lost favor with them. You can imagine the couple's concern when they find that the dog occasionally growls at the baby. Fearing that the dog may snap, they remove the dog from the room when the baby is present. Thinking that the dog may be "jealous," however, they shower attention on the dog, more than ever before. The important point is that the affection is given only when the baby is not around; when the baby is present, the affection is terminated, and the dog may be put outside.

The dog naturally develops a dislike for the child. When it sees the baby, the dog may display its dislike by threatening the baby or acting aggressive. This makes the parents all the more wary and, as time goes on, the contrast in the affection the dog gets when the baby is and is not present becomes greater. One wonders if, in the dog's mind, it does not feel that were it not for the baby, it would get affection all day long. In discussing the problem, clients may reveal that they have tried punishing the dog for threats to the baby. The threat of punishment may force the dog not to display overt aggressive behavior when the parents are present, but it is not going to make the dog like the baby. In fact, the presence of the baby is likely to signify to the dog that it is apt to be punished.

Therapeutic Approach. This is similar to the indirect approach to aggressive dominance in that the rewards of affection and attention are manipulated to get the desired effect. The owners should completely withdraw all attention and affection from the dog when the baby is not present. When the baby is present, one or both parents should give the dog a great deal of affection, attention, and even highly favored food tidbits. At all times, the dog's attention should be directed toward the baby. The idea is not to try to sneak the baby up on the dog. Since it may be difficult for owners to withdraw from the dog as prescribed, it is usually necessary to take the time to explain the theoretical background of this approach. You might suggest they experiment for a week or two with your suggestions. If the dog has an aggressive history with the baby, it will be necessary to protect the baby in the early phases of treatment, using a highchair, a backpack, or other safe enclosure.

Controlling the Baby. The problem with a dog disliking the baby is not only a function of the owners giving affection and attention at the wrong time, but also the tendency for babies to be less than gentle when they contact a dog. Some children are old enough to be taught to approach and pet dogs gently.

Very small toddlers are a different matter. At the risk of alienating some clients, one might suggest that it is not entirely unreasonable to use some behavior therapy on the toddlers. To condition a baby not to crawl toward a dog, for example, one might use a mild water spray from a water gun or plant sprayer when the toddler approaches the dog. The baby will then become more wary of approaching the dog. The water sprayer should be used as a mild remote punishment, and should be delivered so that the baby does not know who sprayed it. In this way the baby's punishment is directly associated with the dog.

Controlling the Dog. The principles used in resolving problems of sibling rivalry can also be used in preventing them. Advice is sometimes sought before the problem emerges. For example, a couple who is expecting a new baby may hear of this problem and consult a practitioner.

When advising people about preventing this problem, focus on the most desirable goal for the behavior of the dog. We are not looking for the dog to grudgingly tolerate a baby or newcomer, but to love it.

The attention the dog normally gets throughout the day should be curtailed, but affection and attention should be given freely to the dog when the baby is in its presence. Thus, the baby bocomes a stimulus that signals the occurrence of affection and attention, and the dog then comes to welcome the appearance of the baby. The greater the contrast in the availability of affection when the baby is present and the absence of affection when the baby is gone, the more readily the dog will learn to like the baby.

In dealing with aggression directed toward people, we see the value of manipulating affection and attention to alter the behavioral and emotional responses of the dog. For some clients, to withhold and deliver affection on a prescribed basis may be very difficult. It may be useful to explain that it is only for a short period, and that there are sound theoretical reasons for using this approach. One might emphasize that this is only a temporary arrangement, and that the owners can expect to return to their normal way of interacting with the dog once the problem is resolved.

When clients are able to manipulate affection for their dog, dramatic results are possible. Sammy's owners (Case 3–4), concerned for their grandchildren's safety during periodic visits, exploited Sammy's dependence on affection. Visits by the grandchildren were structured as training sessions, when Sammy could only receive attention and food treats from the grandchildren, not from the owners.

In some cases, a child actively provokes a dog, and the dog reacts too strongly for the child's safety. It can then be worthwhile to condition the dog to like the poking, handling, or hair pulling that sometimes occurs with small children. Such a procedure was fruitful with Plucky (Case 3–5) when baby Janey persistently pulled lightly at his hair. Although hair pulling is undesirable, with a 2-year-old it is probably unavoidable.

CASE 3–4

History. Sammy, a 3-year-old Rhodesian Ridgeback, has always been friendly toward adults and most children and is dearly loved by the owners who are grandparents of two young children. The dog got along well with the grandchildren until Christmas. At this time, with the grandchildren around and receiving a lot of attention, the dog started growling at the children when they approached him. The situation has not improved and the grandparents, who are still very attached to the dog, are concerned about the safety of the grandchildren. The dog's behavior is still very friendly in the yard in front of the house.

Diagnosis. Competitive Aggression Toward Children

General Evaluation. The goal is to condition Sammy to "love" the grandchildren. The owners are to withdraw all affection from Sammy several hours prior to the grandchildren's visit.

SPECIFIC INSTRUCTIONS

1. Beginning five hours before the grandchildren's visit, give Sammy no affection or treats. Ignore him totally.
2. Instruct the grandchildren to praise and love Sammy when they arrive, even to slip him food treats when they first get out of the car. The children can then spend some time with Sammy playing in the yard, where he has not growled at them.
3. Inside the house, continue with the children as the sole source of attention for Sammy, while also watching him carefully for aggressive behavior.
4. Once the problem is resolved, the grandparents can again give Sammy the usual affection.
5. Progress check in 2 weeks.

The conditioning was to protect Janey, and not to encourage Janey to pull Plucky's hair.

Fear-Related Aggression

In fear-induced aggressive behavior, the aggression is often directed toward specific people or types of people, such as children, adult men, people of a particular race, people in certain uniforms, and so forth. Sometimes it is just a fear of all strange people. There may have been an incident in the dog's past when it was hurt by someone, and the dog then developed a fear of anyone resembling that person. A dog may have been mistreated by a man, for example, and then developed a fear of all men. To cope with its fear, the dog drives people away by growling and snapping when they approach. When growling does not drive people away, the dog often resorts to snapping. If, for some reason, snapping ceases to be effective, the dog may resort to serious biting. Conditioning can shift the level of aggressiveness to more intense forms with time.

History. The diagnosis of fear-induced aggressive behavior is usually obvious from the client's description and the observation of the dog's behavior in the examination room. Characteristically, the dog will leave

CASE 3–5

History. Plucky, a 2-year-old Samoyed, was an agreeable pet until baby Janey began walking. More than anything, Janey enjoyed lightly pulling on Plucky's hair. Plucky responded by snapping and growling, and generally over time becoming more and more aggressive, particularly with Janey.

Diagnosis. Competitive Aggression Toward Children

General Evaluation. The owners will condition Plucky to enjoy having his hair gently pulled. Later, Janey will be present during hair-pulling, and eventually she will do it herself. Meanwhile Janey should be rewarded with affection and food treats to pet Plucky nicely.

SPECIFIC INSTRUCTIONS

1. Begin by patting Plucky and giving her a favored food-treat reward. Use 10 trials per session. Gradually increase the pressure stimulus until mild hair pulling is introduced. Conduct trials on a special throw rug in the same room each day. Then move up to having another adult conduct the trials, continuing the trials until hair pulling is liked.
2. Introduce Janey by having her in view while Plucky's hair is being pulled.
3. Finally, have Janey be the one who pulls the hair.
4. Give a food treat after each trial initially, and then taper off.
5. Be sure to supervise the interaction between Janey and Plucky to ensure Janey's safety.

or attempt to escape from the person with whom it acts aggressive. If it has a fear of men, this may be evident even in the examination room when a man slowly approaches the dog and the dog moves away. Before the threat, the dog may even hide under the chair the client is sitting on. The dog's facial posture is one in which the ears are down and the head is held low. The tail is usually down also.

Therapeutic Approach. Once it is evident that we are dealing primarily with fear-induced aggression, rather than a desire by the dog to be dominant, it is obvious that punishment cannot cure this type of aggressive behavior. When the owner forces a dog to hold still while the fear evoking person approaches, the dog's emotional response will be more intense the next time.

The first step in treating this problem, as with any phobia or fear reaction, is to institute a program that includes systematic desensitization (see Chapter 22). Concomitant with desensitization, it is also desirable to employ counterconditioning by giving the dog a food treat in the presence of a mild form of the fear-evoking stimulus. The specific systematic desensitization program involves gradually exposing the dog to the fear-evoking person while presenting the dog with rewards, especially tasty food, to create a desirable emotional state. Examples of appropriate food treats are small bits of flavorful cheese or hot dogs, not dog biscuits, which are rather bland. The best way to use affection and attention for rewards is for the owners to withdraw affection from the dog except during training sessions when affection is used for a reward.

Exposures to a fear-inducing person should be scheduled as training sessions. Using the example of fear of children, one might have the problem dog sit in a room and a child simply appear in the dog's visual field about 20 feet way. The dog's attention should be directed toward the child, and assuming the dog's behavior is neutral (showing no fear signs), the dog is petted and given a food reward. One appearance of the child followed by one food reward constitutes one trial. These trials combine systematic desensitization (distance gradient) and counterconditioning (food reward). A session might consist of 10 trials. At least one or two sessions can be conducted per day. Over a period of days the sessions should involve bringing the child closer and closer to the dog. Eventually the child can give the dog a food treat or be the source of affection. During the weeks when training sessions are being conducted, every attempt should be made to prevent the dog from being exposed to fear-inducing people outside of the training sessions. This could cause serious regression of the training.

Assuming that the training sessions proceed favorably, and one fear-evoking person is able to approach and handle the dog without provoking an emotional reaction, other people should be systematically run through the training sessions as well. From this standpoint, it is best to start the training sessions with the person least likely to evoke fear reactions. With children this would be the child the dog is most familar with. Sessions with other children should go faster than the initial set of sessions.

Sometimes fear-related aggression presents a particular problem. In the case of Jaws (Case 3–6), for example, the owner sought help because the dog was not performing acceptably in the show ring. The aggression posed no problem for the owner elsewhere, as Jaws was normally only exposed to the family and close friends. The owner was highly motivated to prepare Jaws for the show ring. She faithfully conducted the recommended series of training sessions, and returned for two follow-up appointments to refine the training process. Jaws showed small setbacks when unpredictable factors intruded in the training sessions, but the general pattern was rapid and consistent improvement.

If it seems almost impossible to allow any fear-evoking person to come into the dog's visual field without evoking a fear reaction, the use of a tranquilizer may be necessary. Either a benzodiazepine or a phenothiazine drug may work (see Chapter 24). A behaviorally effective level should be obtained which blocks the fear response but does not produce ataxia or excessive sedation. The dog must be maintained continuously on the tranquilizer (three times per day dosage) until it is well into the training session. As the sessions are continued, the amount of tranquilizer given in each administered dose can be gradually reduced. A sample

CASE 3–6

History. Jaws was a mild-mannered Rhodesian Ridgeback until he was boarded at the owner's daughter's home at 1½ years of age. For some unknown reason the dog became very apprehensive of strange men and threatened to bite judges in the show ring. This aggressive behavior was apparent in the examination room, where the dog withdrew and threatened when approached by a strange male. The owner's goal is for Jaws to be comfortable in the show ring.

Diagnosis. Fear-Related Aggression Toward Men

General Evaluation. The object of the training sessions will be to desensitize Jaw's fear and condition him to *like* being handled by strange men.

SPECIFIC INSTRUCTIONS

1. Conduct a graded series of training sessions with 10 trials per day at home. Rehearse the show-ring routine, using a small rug for Jaws to stand on. Give a food reward for good behavior at the end of each trial.
   a. Have a family member play the role of judge for 5 days.
   b. Have male friends play the role of judge for 5 days.
   c. Have male strangers play the role of judge for 5 days.
2. Conduct training sessions in a practice ring, using same routine.
   a. Have a family member play the role of judge for 2 days.
   b. Have male friends play the role of judge for 2 days.
   c. Have male strangers play the role of judge for 3 days.
   d. It is critical that the training sessions not provoke excessive fear. If Jaws is fearful, use a smaller gradient.
3. Conduct training sessions at a dog show, on the sidelines, using the same routine.
   a. Have a family member play the role of judge.
   b. Have male friends play the role of judge.
   c. Have male strangers play the role of judge.
   d. Phase out the use of the small rug by gradually moving it away.
4. Enter a real dog show.

reduction schedule might be to lower the daily dose by 10 to 20% every other day while the sessions are continued. If the dog reacts adversely, the dose must be adjusted upward again for a longer period until desensitization is achieved.

Pain-Induced Aggression

In general, this type of aggression occurs as a response to painful stimulation. Self-protection against pain is the natural inclination of both animals and people. In the clinical realm, we are most likely to see this type of aggression when a dog suffers from localized pain and then snaps or growls if handled near the painful area.

History. The most common signal of pain-induced aggression is when a dog acts aggressive only when a specific region of the body is touched or manipulated. If the existence of a traumatizing foreign body or inflammatory process is not obvious, the usual approach of using radiographs and other diagnostic techniques to reveal the cause of the pain are then in order. After the cause of the pain has been resolved medically, and a dog still acts aggressively when touched or handled in the area, the behavior can be viewed as a conditioned response which is amenable to behavioral therapy.

Therapeutic Approach. It is, of course, necessary to remove the source of the pain before expecting much progress with this type of aggressive behavior. Once it seems as though the pain is no longer occurring, the indicated treatment is systematic desensitization plus counterconditioning.

The desensitization gradient could begin by lightly touching the dog directly over the area, and giving food treats each time this is done. After a couple of daily sessions of 10 trials per session, the owners can be more heavy handed in touching the affected areas until they are eventually using normal pressure. The total conditioning time might be two weeks if the process is continued on a daily basis. Gradually touching the dog closer and closer to the affected area while normal petting pressure is used is another desensitization gradient. For a sensitive area near the back leg, for example, one might pet the dog's back and give a food treat and affection after each petting sequence. Then the petting could gradually be transferred toward the direction of the affected area on the leg.

In some cases, the desensitization may be more expediently approached with a tranquilizer. If the conditioned pain response is eliminated by the drug, the owners should handle the dog in the affected area frequently, rewarding the dog with praise, affection, and food treats. The dosage of tranquilizer is then gradually lowered over the next two weeks.

The similarities in treatment approaches between pain-induced aggression and fear-induced aggression are quite obvious. Both require a systematic desensitization process and counterconditioning of the anxiety reaction underlying the fear. Desensitization is accomplished by using a gradient of stimulus strength ranging from an initial safe level to the level of full stimulus strength. In both instances, tranquilizers may be employed to reduce emotional reactions and allow a more rapid desensitization process. Gradually phasing out the tranquilizer after several days of stimuli exposure is also a type of gradient.

Territorial Aggression Toward People

It is possible to have too much of a good thing, and this is occasionally true of the territorial guarding behavior of dogs. Guarding can be so extreme that it keeps friends and relatives out of a yard or house.

History. A dog may be perfectly friendly toward people away from home, but act fiercely aggressive toward those same people when they are in its territory. Sometimes only strangers are threatened, such as mail delivery people and meter readers. Guarding behavior is reinforced because a dog always chases away the intruders. The persuasiveness of this behavior is due to the dog's natural tendency to protect the territory of the pack. Occasionally, owners have rewarded or praised the dog for aggressive guarding and this has reinforced the behavior.

Therapeutic Approach. If the dog has no fear of the people it is aggressive toward, then there are two options for dealing with the problem. One is to bring the dog's behavior under voice and hand control by punishing the aggressive behavior whenever it continues after the dog has been told to be quiet or sit. The dog is likely to still be aggressive when the owners are not around to discipline it. This, of course, is just what some people want.

The second approach involves attempting to condition the dog to like the people who come to visit. Basically, the dog is induced to change its attitude toward intruders by rewards. A technique that has proven successful is for the members of the dog's family to ignore the dog and not give it any attention or affection except when visitors come over. Only when a visitor is present does the dog receive affection. At these visitation times the visitors, and later the owners, can give the dog as much attention as they like. Because the dog receives the affection it desires only when visitors are present, it will come to look forward to visitors entering its territory and not want to chase them away.

In some cases both options can be combined, punishing the aggressive behavior and also conditioning the dog to like visitors. A combinational approach of this type was used with Lobo (Case 3–7), who was initially punished for aggressiveness with strangers until she gradually became more controllable. Then affection withdrawal by the owners was begun, and social contact was only provided by strangers.

The procedure of conditioning the dog to like people who come to visit is very similar to that recommended for sibling rivalry discussed in the preceding section. It may be a good idea for the owners to arrange a succession of visits from people that the dog is most familiar with, and then progress to people that the dog is less familiar with. The dog should understand the significance of a visitor's presence after a few sessions with the familiar visitors.

A variant of this technique can be used for aggression toward specific people such as mail carriers. A series of trials should be set up each day in which the uniformed intruder approaches the dog on its territory from some distance, but stops short of a point where the dog becomes aggressive. The owner, standing beside the dog, should reward it with food when the intruder reaches a certain point. Over a series of daily sessions of about 10 trials per session, the intruder gradually gets closer to the dog. The goal is for the visitor to pet or handle the dog before the dog is paid off. The most successful outcome of this reward-based

CASE 3–7

History. Lobo, a young female Labrador Retriever, has become very aggressive toward anyone but family members that enter her territory. Her territory is a large orchard. She has been prevented from attacking several people by last minute heroic efforts by the owners. These clients seem willing to invest considerable time to change this aggressive behavior.

Diagnosis. Territorial Aggression Toward People

General Evaluation. Lobo will be severely punished for showing aggressiveness toward strangers. Once she is more controllable, she will be conditioned to like people who come to visit.

SPECIFIC INSTRUCTIONS

1. Stage episodes that allow severe punishment to be safely administered to Lobo when she shows aggressiveness toward strangers. Have a stranger walk by the territorial border while Lobo is tied securely. Then, when she is aggressive, yell and hit or slap her immediately. After an incident, withhold affection for several hours.
2. As Lobo becomes controllable, condition her to like strangers by withholding all affection from her. Your withholding of affection should be complete and total, including not allowing her to sleep in the bedroom or house. Try an affection withdrawal approach for a maximum of 2 weeks to determine if this approach will work. Strangers are to become the only source of social contact. Begin with acquaintances she tolerates as a source of affection, and gradually switch to strangers. If Lobo begins to show aggression during these episodes, isolate her and totally ignore her until the next episode.

approach is a dog that loves territorial intruders. To some clients this outcome may be as undesirable as the overly aggressive territorial guarding behavior. One cannot necessarily have it both ways. An intermediate level of behavior change might be possible, but this type of outcome would be difficult to program.

The techniques for dealing with territorial aggression work most effectively if the dog is first trained in an area near the borders of the property where territorial aggression is weakest. If the front yard is less risky than the back yard, for example, initial phases of training should be in the front yard with a goal of working toward the backyard.

Idiopathic Vicious Attacks on People

This type of aggressive behavior is characterized by unpredictable and unprovoked vicious attacks on people the dog knows well. The attacks are infrequent, often spaced a month or more apart. There is evidence of a genetic predisposition toward this behavior, and also of subclinical inflammation in the brain.

Anyone who proposes to treat this behavior is faced with the problem of interpreting the effectiveness of the treatment, since the episodes of aggressive behavior may be only at monthly intervals. How do you know if the drug, conditioning program, or other treatment is effective if you have to wait until you feel an attack would likely have occurred by chance? Since the dogs are invariably well-behaved between attacks, one cannot monitor the effectiveness of treatment on a day-to-day basis as with other types of aggressive behavior. Because the occurrence of aggression is unexplainable, and treatment extremely difficult to interpret, this is the one type of aggressive behavior for which euthanasia is clearly indicated.

History. The typical case is the dog that the owner describes as usually friendly, affectionate, and well-mannered. The owners are almost dumbfounded by the attacks in which the dog, for no explainable reason, suddenly turns and viciously attacks a member of the household or a friend of the family. The behavior is clearly unprovoked and unpredictable. Usually the dog gives little or no warning before an attack. The attacks may be directed toward a person's face, neck, or arm.

Owners of such dogs are likely to mention that moments before the attack the dog no longer seems to recognize people in the family, and may even get a glazed or distant look in its eyes. After the attack some dogs appear subdued while others act as though they were not aware of what they had done. Among the breeds in which this behavior has become a noticeable problem are St. Bernards, Doberman Pinschers, Bernese Mountain Dogs, and German Shepherds.

The attacks initially seem to occur sporadically, perhaps about a month apart. When they become more frequent, professional advice is often sought. A triggering stimulus that is sometimes mentioned is when the dog is given a command, in a very calm or even friendly manner. Such commands would normally be no cause for any type of aggression.

Dogs with this aggressive syndrome do not usually display any evidence of clinical abnormality. In some cases, where the animals have been subjected to euthanasia, examination of the brain has revealed no pronounced pathology of the nervous system or other organ systems. A mild degree of encephalitis has been observed upon careful microscopic examination of parts of the brain in some dogs (Hart, 1977). With painstaking neuropathological examination, perhaps more could be documented in the way of underlying brain pathology.

There is evidently a genetic predisposition toward this behavior. A survey of Bernese Mountain Dogs by Van der Velden and coworkers (1976) traced the occurrence of this behavior in dogs in Holland to two males imported into the country.

Therapeutic Approach. Probably the best advice to clients, in the interest of their own safety and that of others, is that the dogs be euthanized. Some owners, particularly those of small dogs, may insist that treatment be attempted. On the basis that this form of aggressive behavior may reflect abnormal eruption of neuronal activity, some clinicians have

CASE 3–8

History. A lifelong pattern of chasing and snapping at strangers and a resistance to being dominated by his owners has been typical of Morris, a 2½ year old Great Pyrenees male. Now a new problem has been added: sudden vicious attacks on people the dog knows well. Morris has had to be pulled back from unprovoked attacks on family members, and at these times he is uncontrollable.

Diagnosis. Idiopathic Vicious Attacks

General Evaluation. Morris exhibits two types of aggressive behavior, dominance-related aggression with people and idiopathic vicious attacks.

SPECIFIC INSTRUCTIONS

1. To correct aggressive dominance, we would usually recommend obedience training, and indirect or direct control of the dominance.
2. For idiopathic vicious attacks, treatments are not reliable. A dog of this size poses a great risk to the people around it, and there is a history of similar behavior in the dog's grandfather. Since attacks are sporadic, there is no way of knowing when another will occur, or with whom. Thus, the success of a treatment cannot be evaluated. The owner is likely to be a target sooner or later. Euthanasia must be recommended.

found that anticonvulsant drugs such as primidone, phenobarbital, and diphenylhydantoin are useful in controlling the attacks. Of course, the problem with attempting to treat the behavior is that the attacks occur infrequently and are not predictable. One may have to treat a dog for several weeks or months and not know if the treatment is effective or if the dosage should be adjusted until another attack occurs. Several people could be attacked while the treatment is being evaluated. These dilemmas were all too apparent in the case of Morris (Case 3–8), a Great Pyrenees, who regularly contested for dominance, and who also exhibited violent attacks on family members that were totally unprovoked, unpredictable, and vicious.

TREATING AGGRESSIVE BEHAVIOR DIRECTED TOWARD OTHER DOGS

It is perhaps somewhat surprising that there is little overlap in the instances of aggressive behavior directed toward other dogs and that directed toward people. Dogs that are aggressive with other dogs are usually friendly toward people.

Aside from the occasional occurrence of predatory aggression, where a large dog may attack a small dog, there are two causes of fighting between dogs. One is the failure of the dogs to establish or maintain a peacekeeping dominant-subordinate relationship. The other is the innate tendency of dogs, particularly strangers, to pick fights with members of their own sex. The latter problem is especially prominent in males and is referred to as intermale aggression. Occasionally, one finds females that act aggressively only toward other females. This type of aggression might be referred to as interfemale aggression. If fighting behavior occurs only when a bitch is in estrus, the behavior may be due to secretion of estrogen. Ordinarily, fighting between females can be attributed to a problem in developing a successful dominant-subordinate relationship.

Aggressive behavior between strange dogs may be a manifestation of the territorial protectiveness of dogs. This problem should be handled by direct control, much as one would handle territorial aggression toward people. Of course, when a new dog is introduced into another dog's territory permanently, a dominant-subordinate relationship must be established if the dogs are to get along without fighting. There are instances, however, especially among the less aggressive breeds, where parallel relationships can be maintained and the dogs do not fight (Solarz, 1970).

In most instances where two or more dogs live together, there will be a dominance hierarchy. Dominance is often determined by size, although other factors may enter the picture. Males tend to be dominant over females; but this is not always true. The animal that has lived in the territory the longest is often the dominant one, even though it may be smaller. Dominance is expressed and reinforced by facial expressions, certain body positions, and eye contact. A subordinate often acknowledges the signs of dominance by diverting eye contact with the dominant animal, turning its head to the side, lowering the tail, and moving away. When necessary, dominant dogs can intensify the social signals involved in dominance by growling, snarling, or even snapping at subordinates. These interactions may be seen at the time of feeding or when both animals interact with the owners. The social signals that are displayed by both the dominant and subordinate animals may be so subtle that the owners do not notice. Sometimes owners are not aware of which dog is dominant.

The dominance hierarchy is a dynamic relationship that is subject to alteration. This is seen, for example, when a family has a small dog that has been with them for a long time, and they decide to get a second dog. Their family favorite, a Dachshund, for example, may be getting old, and they want a puppy as a companion for the older dog or as a potential replacement. Without giving the matter much thought, they may decide that the new dog should be a larger breed, so they get a German Shepherd. Initially, the old family favorite will be dominant over the puppy, but as time goes by and the puppy begins to mature, it is going to have a major size advantage over the Dachshund. The dominance reversal could come about slowly. In other instances, the dominance reversal could be triggered suddenly over possession of a bone, and one good fight will settle the new dominant-subordinate relationship and put the larger dog in charge. This reversal of dominance may occur without the owners realizing that it has happened.

Another way in which reversal of dominance occurs is when people adopt a new adult dog that is larger than the resident dog. The new dog, lacking seniority, may be happy with a subordinate role at first, but later, it will challenge the first resident for its position. Again, the dominance reversal may occur without the owners knowing it.

Since fighting brings with it the possibility of injury to one or more dogs, our goal is to help establish or maintain a dominance relationship that keeps peace among the dogs. This might be accomplished by physically punishing the dog destined to be the subordinate whenever it is observed standing up to the dog destined to be dominant. In fact, when the dominant-to-be dog growls or threatens, it can be assumed that it has reason to threaten, and the subordinate could be punished whether it looks like it deserved to be or not.

One approach for a severe fighting problem that appears to reflect unsettled dominance is to continuously administer each dog a tranquilizer and thus reduce the emotional responses that lead to fighting. For each dog, a maintenance dose is determined that allows the dog to move about, eat, and interact with people, but blocks the tendency to fight. Both dogs should be on a maintenance dose for one week. This should make it easy for the dominant dog to reinforce its position, so that after about two weeks, both dogs can then be gradually taken off the tranquilizer. As long as the owners help reinforce the dominance relationship, the tranquilizer-facilitated approach should ease the dogs into a stable relationship without a series of severe fights.

Dominance-Status Aggression

With the above background on the importance of dominant-subordinate relationships in keeping peace among dogs, we turn to a common problem.

History. Fighting between dogs is often provoked or stimulated by the way the owners interact with the dogs. This leads to the classic situation where severe fighting between two dogs breaks out when the owners are around, but never in their absence. This bit of history is sufficient evidence that the main cause of the fighting is the owner's interference with the dominant-subordinate relationship that normally keeps peace among the dogs when the owners are gone.

One way to look at this problem is that when the dogs are around the owners, the most desirable goal for both dogs is gaining attention from them. The dominant dog expects to be the first one to receive attention and it expects the most attention. The subordinate, having an equally strong desire to obtain affection from the owners, may try to horn in on the dominant. The dominant dog may then be compelled to reinforce its position by threatening or snapping.

Our tendency to favor the underdog adds to the problem. We want to punish the instigator of the aggression, which is usually the dominant animal. Soon the subordinate realizes that in the presence of the owners it is protected from the dominant animal. The subordinate then tends to stand up to the dominant for attention from the owners. This insubordination can evoke a full-blown aggressive attack from the dominant dog. To make things worse, the owner not only may punish the dominant animal, but also remove it from the house. This, of course, makes the dominant dog even more resentful and aggressive toward the subordinate.

This type of aggressive behavior is common when a family gets a puppy of a larger breed to be the companion of the old family favorite of a smaller breed. The dogs may have long since reversed their dominance relationships, but the smaller dog may know that in the presence of the owners it will be protected and, therefore, it will continue to stand up to the larger dog when the owners are around. It is only under these circumstances that the larger dog attacks the smaller one.

Therapeutic Approach. To treat this problem, it is important to make the owners understand that social relationships between dogs cannot be handled in the same way as human social relationships. It can be pointed out, for example, that it is natural for dogs to accept a dominant or subordinate role, and it is natural for the dominant dog to regularly reinforce its position with gestures, threats, or physical punishment, and for the subordinate to respond by submitting to these gestures.

The therapeutic approach involves determining which dog is dominant if this is not already clear. The owners may have to experiment, observing the dogs when they are not aware of being watched. The animal that gets the bone, gets in the automobile first, or seems to control the best resting spots is probably the dominant one.

The owners must then treat the dominant animal as the primary dog when both are greeted after an absence. The dominant dog should be treated with all the respect and privileges that accrue the top dog. This includes receiving attention first and more frequently. When the dogs are taken for a walk, the dominant dog should be attached to the leash and allowed outside first.

The subordinate, because the owners have intervened in the past, may still try to push the dominant aside to obtain the owner's attention during greeting and at other times of interaction with the owner. The subordinate should then be punished or isolated so that it no longer feels protected in the owner's presence. In short, the subordinate dog must learn that it has to respond to the dominant dog's signals of dominance whether the owners are present or absent. Since the dogs get along when the owners are not around, they have obviously developed a system that works.

The problem of dominance-aggression arose for the owners of Duke

CASE 3–9

History. A couple brought in their two dogs that were engaging in intermale aggression. Duke, a Chesapeake Bay Retriever, and Prince, a Golden Retriever, were habitually fighting, and Prince had sustained severe injuries. Both dogs were gonadally intact. Prince, a rather timid and withdrawn dog, had been the woman's favorite for years and enjoyed special privileges of spending the day in the house and sleeping in the bedroom; whereas Duke had been relegated to the garage since the difficulty. The dogs apparently did not fight when the couple was away from the home.

Diagnosis. Dominance Status Aggression

General Evaluation. Although the owners prefer and support Prince, the Golden Retriever, the dogs know that Duke, the Chesapeake Bay Retriever, is dominant. Yet Duke is not accorded his rightful status.

SPECIFIC INSTRUCTIONS

1. Reinforce Prince's position as the subordinate dog. Stop letting Prince up on the couch, and don't let him sleep inside while Duke is outside. Don't give Prince privileges that Duke doesn't get.
2. When Duke growls, assume that Prince did something wrong. Punish Prince by scolding him, and then put both dogs outside.
3. Treat Duke as the dominant dog. Feed and greet him first, and give him more attention than Prince gets. These instructions are particularly crucial for the woman owner, who previously preferred Prince.
4. For the first week or two, really overdo the preferential treatment for Duke. This will emphasize the new relationship between yourselves and the dogs.
5. If the problem persists, castration of Prince is recommended. This should reduce aggression in Prince and make him a less appropriate target for Duke to attack.

and Prince (Case 3–9), who sought help when Duke persisted in attacking Prince. As a result of the attacks, the owners had compounded the problem by withdrawing many of Duke's privileges, considering him the "bad dog." For treatment, the owners reversed their patterns of interactions with the two dogs, showing a consistent preference for Duke and assuming that any problems were instigated by Prince.

Keeping dogs apart when they have this problem does not ameliorate the difficulty. Indeed, some interaction is required to begin solving the problem. But the question, of course, is how to put the dogs together without a mishap. In extreme cases, a muzzle or tranquilization can aid in the initial period of resocialization of the dogs with each other, as was done with Casy, Rebel, and Manny (Case 3–10).

Favoring the dominant dog is contradictory to our natural tendency to punish the aggressor. Your advice may not be readily accepted, and this is why it is necessary to explain the theoretical basis of the recommendations.

Finally, a bit of advice to the owners may be to suggest that they "play

CASE 3–10

History. Two castrated Springer Spaniel brothers, Casy and Rebel, are fighting often, and sometimes the intact male Golden Retriever, Manny, joins in. In the past Manny has been dominant. The owner says that now Casy is becoming dominant and, in support of this, Casy usually initates the aggression. Recently, Casy has been isolated from the other dogs.

Diagnosis. Dominance-Status Aggression

General Evaluation. The approach here is to establish and support the dominance hierarchy of the dogs, with the initial aid of tranquilization.

SPECIFIC INSTRUCTIONS

1. All three dogs will be tranquilized for one week. Meanwhile, support Casy, the dominant dog, and punish the other two dogs for any infraction of the dominance hierarchy. After one week start phasing out the tranquilizers while continuing to support the dominance hierarchy.
2. Castrating Manny, if you chose, may reduce his aggression.

down" their greeting responses, so that both dogs will be less excited about competing for the owner's attention.

If fighting between the dogs occurs in the owner's absence as well as their presence, the problem may be an unresolved dominant-subordinate relationship. In addition to using behavioral techniques to establish a permanent dominant-subordinate relationship, the practitioner might treat one of the dogs with a tranquilizer. Tranquilizing just one dog can make it easier for the other dog to obtain a dominant position. If the dogs that are fighting are both gonadally intact males, some thought might be given to castrating the one determined to be the subordinate by virtue of size or seniority.

Fighting Between Strange Male Dogs

History. Most dogs are either indifferent to strange dogs or quick to interact with them and settle into at least a temporary dominant-subordinate relationship. Some dogs immediately react to strange dogs by fighting.

Therapeutic Approach. Castration is the indicated treatment for this type of aggression. Since this is the only type of aggressive behavior for which this operation is indicated, it is important to establish the diagnosis. Clinical experience indicates that castration eliminates, or markedly reduces, the tendency of male dogs to engage in fights with other males about 60% of the time (Hopkins, *et al.*, 1976). The main effect probably results from the loss of testosterone, removing its effects upon the neural systems in the brain that mediate intermale aggression. Another effect of castration is to cause a change in the odor of male dogs, making them a less provoking stimulus for other males.

When castration is not effective in resolving intermale aggressiveness, the administration of a progestin is indicated. The injection of one dose of medroxyprogesterone at the rate of 5 mg per kg, or the oral treatment with megestrol acetate at the rate of 0.5 mg per kg daily initially, as outlined in Chapter 23, may be effective (Hart, 1981a). Clinical experience points to a success rate of about 75% with progestin treatment in castrated males.

When two male dogs residing in the same household are fighting, and both are gonadally intact, one way to assist in the development of a successful dominant-subordinate relationship would be to castrate the dog that appears to be the best candidate for the subordinate position. This procedure has been successful in a number of such instances, undoubtedly, because the castrated dog has less tendency to fight and stand up to the other dog, and also because the odor change makes him a weaker stimulus to provoke aggression. If the above procedure does not work in the course of a month or two, castration of the other dog should be considered to weaken its tendency to fight. Behavioral procedures should still be followed to enhance the dominant dog's position as outlined in the preceding section.

Another possibility for solving the problem of fighting between strange dogs is counterconditioning, and it has applications for dogs of either sex. Staged encounters can be planned when a friend brings another dog within view of the problem dog, but at a distance great enough not to evoke an attack. The problem dog is commanded to sit, but its attention is directed to the other dog. It is given a food treat when it sees the other dog. During each session several trials are conducted with the friend's dog being brought into the problem dog's view and the problem dog given the food reward. Over a series of daily sessions the friend's dog is brought closer and closer. Eventually the problem dog may change its attitude toward the friend's dog. Another stimulus dog can be introduced, again at a distance, and the process repeated. The counterconditioning, if consistently pursued, may change the problem dog's attitude toward all dogs.

Chapter 4

Fear and Emotional Reactions in Dogs

To be fearful of a potentially harmful situation is normal and healthy for animals and people, but excessive fear is counterproductive. Our understanding of fear behavior in dogs is aided by our realization of what it would be like for us if other people were not able to talk to us about our fears or explain that strange things will not hurt us. Aside from explanations that something will not hurt us, we lose fears of threatening stimuli such as thunder or firecrackers, when, through the process of habituation, we repeatedly experience these stimuli and no harm befalls us. It is only through such direct habituating experiences that animals are able to adapt to sounds that are otherwise fear-evoking. No amount of assurance or verbalizing will explain to a dog that loud sounds are harmless. Good gun-dog trainers recognize the importance of adapting dogs, as very young puppies, to potentially frightening stimuli. Hence, puppies are habituated to gunshots early. Puppies adapt to these stimuli more easily than adult dogs. Problem cases involving fear are common in adult dogs, and the fear has often progressed to the point where the dog may be destructive in attempting to escape from the frightening stimulus.

THE NATURE OF FEAR REACTIONS

Types of Fear Reactions

Dogs can experience extreme fear reactions to inanimate stimuli such as thunder, lightning, fireworks, or gunshots. Often the owners offer a great deal of comfort and affection to their dogs when they are particularly fearful. This comfort can enhance fear reactions, because the dog's fear behavior is reinforced.

Sometimes dogs' fears relate to other dogs or people. This links fear reactions with the social responses of dogs toward other dogs or people. They may be aggressive in an attempt to drive off a fear-evoking person.

A fear reaction in response to being left alone, commonly called "separation anxiety," is a frequent behavioral problem in dogs when owners start working full-time after they had previously provided constant company for a dog. Dogs are social animals, and a fearful emotional response upon being suddenly left alone is an expected reaction. The emotional disturbance is a rather generalized one, and it cannot be directed toward people, other dogs, or specific stimuli. Therefore, a wide range of misbehaviors may be seen, including destructiveness, excessive barking, and inappropriate eliminative behavior.

Causes of Fear Reactions

Fear and other emotional reactions may reflect acquired responses or innate responses that are not habituated. Being shot with a gun is an obvious cause for a dog to develop a fear of fireworks. Being abused by a man could explain a dog's acquired fear of all men. Dogs have a natural tendency to be fearful of intense stimuli and strange experiences. Most puppies live through thunderstorms and fireworks, and without our planning adapt, or desensitize themselves, to the sounds. A puppy that is protected from such sounds and comforted during storms may not adapt to these sounds. The fear reaction in this case reflects a lack of habituation. Dogs that remain quiet and well-behaved when left alone have habituated themselves to the fear of separation.

General Treatment

The treatment designed for fear reactions must correspond to the individual dog's problem behavior. Whether the fear is associated with inanimate stimuli, people, or separation, the stimulus evoking the fear must be precisely identified. To resolve the fear behavior, there must be gradual exposure to the specific fear stimulus.

An acquired emotional response to inanimate objects or people is dealt with by gradually extinguishing the reaction through controlled exposure to the stimuli. Emotional reactions that are nonhabituated responses to intense stimuli or separation are dealt with by controlled habituation. The extinction and the habituation processes are virtually identical in practice, and both are handled through the concept of desensitization, which involves the gradual exposure of the dog to greater and greater intensities of the stimulus. Finding a way of presenting the stimulus in a gradient fashion can be a major task. Various tricks for presenting the stimuli in a graded form are discussed under specific problem cases.

The most successful desensitization programs involve pairing a reward, usually favored food treats, with the lowest gradient of the aversive stimulus. The reward produces a favorable or appetitive emotional state which is incompatible with the undesirable emotional reaction. This is referred to as counterconditioning. As long as the aversive stimulus is fairly weak, the appetitive emotional state can override the aversive reaction and the extinction or habituation process is initiated. The concepts of systematic desensitization and counterconditioning are dealt with in detail in Chapter 22.

Use of Tranquilizers

Tranquilizers may be advantageous in treating fear of loud sounds, particularly if a dog shows an emotional response to the lowest level of fear-inducing stimulus (see Chapter 23). Training sessions should be conducted several times a day under a full dosage of a tranquilizer, administered three times a day. This treatment may produce a desensitization to the low sound level within a couple of weeks. The tranquilizer dosage should then be gradually reduced over the next two weeks while the stimulus is still presented at regular periods. It the dog exhibits desirable behavior once it is off the tranquilizers, then the intensity of the stimulus can be gradually increased. As soon as training sessions can be conducted without the drug, desensitization will occur more rapidly.

FEARS OF INANIMATE OBJECTS OR PARTICULAR SITUATIONS

Practitioners frequently see dogs that are afraid of loud noises, such as thunder, gunshots, or firecrackers. Dogs seem highly predisposed to phobias of this kind, and treatment for some of them will be discussed below. However, dogs can become afraid of a wide range of stimuli, and sound need not be a factor. Systematic desensitization can be adapted to each individual case, regardless of the specific fear object. For example, Suzy (Case 4–1) was fearful of riding in the car. Whether the fear resulted from lack of experience with cars or from a traumatic experience, the treatment would be the same and involve exposing Suzy to a car in a graded series of training sessions.

Thunderstorms

Since most dogs have not been hit by a lightning bolt during a thunderstorm, the fear of thunder probably reflects an innate fear of loud noises. Most dogs become habituated to storms through repeated exposure to them. People who are going to have dogs in places where thunderstorms are common should see that the dogs are exposed to storms as puppies. One might even propose that the owners start exposing the dog to thunderstorm recordings early in life to intentionally induce habituation.

History. Some dogs are so terror-stricken by thunder and lightning that they injure themselves by jumping through screen doors or windows. Dogs frequently approach their owners for comfort during a storm. While it is natural to try to comfort the animal, the affection reinforces the dog's adverse reactions and, therefore, is counterproductive.

Therapeutic Approach. Physical punishment by hitting the dog or punishing it remotely, as with a shock collar (see Chapter 22) is ineffective

CASE 4–1

History. Suzy, a 3-year-old German Shepherd, whines or acts wild whenever the owner takes her in a car. Suzy is a valuable show dog and the owners would like to continue to take her to dog shows. The behavior has gradually developed over the past 16 months.

Diagnosis. Acquired Fear

General Evaluation. Suzy's fear of cars will be gradually desensitized in a series of training sessions.

SPECIFIC INSTRUCTIONS

1. Training sessions will gradually increase Suzy's exposure to the car without arousing fear. Ignore her if she looks frightened during any trial and shift back to the earlier trial level.
2. Schedule 10 trials per day, with three days at each stage of training. For each stage bring Suzy toward the car as follows: within 15 feet of the car; within 10 feet; within 5 feet; touching the car; climbing in the car to stay for one minute; staying in the car while it starts. Continue increasing exposure with rides of increasing length.
3. After each trial give her a food treat. Schedule sessions when Suzy is likely to be hungry.
4. Avoid exposing Suzy to cars except during training sessions.
5. Progress check in 2 weeks.

and will make matters worse by enhancing the emotional response which leads to the behavioral problem. Soothing and reassuring the dog during storms also increases the problem because it tends to reward the dog's emotional reaction. The reaction of the dog's owners to an emotional response following a fear-inducing stimulus should be one of cultivated indifference.

The goal of therapy is to habituate the dog to thunderstorms through desensitization. The basic concept of the thunderstorm desensitization program was introduced by Tuber, Hothersall, and Voith (1974). Using counterconditioning methods, they suggest a series of training sessions to elicit a favorable response, such as being quiet and behaving calmly during thunderstorms. You might think of it as teaching the dog to "enjoy" thunderstorms using appropriate rewards such as food treats and affection. Conditioning begins with the presentation of the sound of a thunderstorm at an amplitude intensity so low that it does not bother the dog. This allows the appetitive emotional reaction produced by the food treats to override the weak aversive reaction produced by the recording.

The owner should obtain a commercial recording of a thunderstorm from a local record dealer. Playing the storm recording once at full amplitude will verify whether it creates the aversive emotional reaction of concern. If it does not, one must locate another recording that elicits the full behavioral response. The effective portion of the thunderstorm and recording may be transferred to tape for repeated use.

Training sessions should be routinely scheduled at a certain time of the day. They, of course, must be conducted where a sound system is available. Training should be conducted at a time of the year when actual thunderstorms will not disrupt the training.

Initially, the dog is taught to sit or lie down in a relaxed fashion for several minutes in the vicinity of the speakers. For the first series of sessions the record is played at a very low volume, a volume so low we know that the dog will not react adversely. About every 30 seconds the dog is given frequent food treats and affection for being good while the record is played at this low volume. After a few sessions the dog should come to enjoy the training with this initial sound intensity.

As the training sessions progress, the owners should begin to gradually increase the amplitude of the recording. The dog should be given approximately twenty trials with rewards per session if its behavior remains relaxed. If the dog shows an obvious emotional response, the training is advancing too rapidly, and the session should be terminated. The owners should return to a previous safe amplitude the next day.

Once the dog has had about 30 to 60 daily training sessions of this type, it may be ready for a full-blown thunderstorm-like level of sound. During either recorded or natural thunderstorms, the owners should continue to reward and praise the dog when its behavior is acceptable.

The treatment will have a better chance of success if it is tailored to the specific situation. In the case of Debbie (Case 4–2), a particular airplane engine intermittently flying over the house elicited and escalated

CASE 4–2

History. The owners of Debbie, a Golden Retriever, live under the flight path of an Air Force base. For some time she has reacted strongly to infrequent thunderstorms. Recently, a new airplane engine is being tested about once a week. Debbie becomes very upset when she hears the noise of this particular plane. She claws and tears the screens in the backyard.

Diagnosis. Noise Phobia (jet airplane and thunderstorms)

General Evaluation. Debbie will be administered a systematic desensitization program from the jet airplane sound that regularly initiates her reaction.

SPECIFIC INSTRUCTIONS

1. Prepare a high-quality recording of the airplane noise.
2. Test the recording at normal amplitude with Debbie present to see if it elicits her reaction.
3. Prepare a sound system with outdoor speakers that are appropriate for outdoor desensitization sessions.
4. When the sound system is ready, schedule a return visit to plan desensitization sessions.

CASE 4–3

History. A reaction to thunderstorms is severely upsetting Buff, a Cockapoo, and his owners.

Diagnosis. Noise Phobia (thunderstorms)

General Evaluation. Buff will be given a systematic desensitization program to eliminate his reaction to thunderstorms. Affection for him will also be controlled.

SPECIFIC INSTRUCTIONS

1. Obtain a thunderstorm record and play it at full volume to see if the emotional reaction is produced.
2. Conduct training sessions with the thunderstorm record at a low volume for 10 minutes per session, at least 2 sessions per day. Reward good behavior frequently (every 30 seconds) with affection and food.
3. Increase the volume of the recording very gradually from session to session.
4. Give no attention to Buff except during training sessions when he performs well. Ignore Buff if he exhibits a fear reaction during a real or artificial thunderstorm.

CASE 4–4

History. Misty's reaction to thunder and firecrackers is increasing. For some time this German Shepherd has been running away, and now, if contained, she breaks windows, screens, and doors in response to loud noises. Her intense reaction includes frothing at the mouth, and even occurs in response to loud noises on TV or radio programs.

Diagnosis. Noise Phobia (thunderstorms and loud noises)

General Evaluation. Misty will be tranquilized to calm her enough that she can participate in training sessions. The desensitization approach will include exposure to a thunderstorm recording on a controlled basis. Misty will be given attention only during these sessions.

SPECIFIC INSTRUCTIONS

1. Give Misty a tranquilizer. Start with diazepam (Valium) at 10 mg twice a day. Increase the dosage 1 tablet at a time. Reduce dosage if the dog has difficulty walking.
2. Begin modified flooding sessions after administering the tranquilizer for a few days, when Misty is noticeably calmer. Play music and TV continually while you are home. Then play the thunder record at the lowest level. Reward good behavior with a food treat. Gradually increase the volume.
3. Give attention to Misty only when the record is on and her behavior is good.
4. Progress check in one week for further instructions.

a fear reaction that was previously a response only to thunderstorms. The training program for Debbie was based on the engine sounds, the primary problem at the time. This necessitated that the owner prepare a recording and sound system similar to the real experience of the jet engine that was heard overhead. In the more straightforward cases, in which thunderstorms are the primary problem, as with Buff (Case 4–3), commercial recordings are available.

Occasionally a dog such as Misty (Case 4–4) reacts so strongly to any loud noise that it is impossible to conduct a training session. Tranquilization is useful for these cases. If the tranquilizer is to be effective it must reduce the fear at a dose below that which would produce ataxia. The typical procedure is to begin with a low dose and increase it for the desired effect (see Chapter 23).

Fireworks and Gun Shots

This type of fear may reflect either an acquired or innate emotional response. However, a diagnosis of the source is not essential, since the therapeutic approach is the same.

History. Dogs are commonly very upset during Fourth of July celebrations. If phobias to fireworks become a problem only one day a year, it may not be worthwhile to desensitize the fear responses. Using a tranquilizer at that time may be sufficient. However, desensitization is necessary for Fourth-of-July celebrations that go on for weeks, or if the dog is regularly exposed to gunshot sounds from hunters, target shooters, and the like.

Therapeutic Approach. The approach for desensitizing dogs to firecrackers or gunshots is the same as for thunderstorms. If a sound recording is used, it must be tested to assure that at its full volumetric intensity the dog demonstates the undesirable behavior.

The simplest and most effective technique for desensitizing a dog to gunshot and firecracker phobias is to use a pistol as the training stimulus. The starter pistols used in track events fire .22 caliber blanks and are the safest guns to use. To create a gradient for the sound stimulus, cardboard boxes of varying sizes can be nested inside one another to muffle the sound. By placing a hole through all the boxes, it is possible to fire the gun from the innermost box while all the other boxes are nested around it (Fig. 4–1). The muffling of the sound must be sufficient so that the aversive emotional reaction disappears with complete muffling. Terry cloth towels can be draped over the boxes for extra muffling. Counterconditioning training sessions can be scheduled with 10 trials or shots for each session. After each shot the dog is rewarded with food. In succeeding sessions a box is removed every three or four sessions until the unmuffled gun can be shot off with no fear response. If necessary, towels can be used between boxes to provide a finer gradient. Training must be set back to the previous gradient if the dog shows an

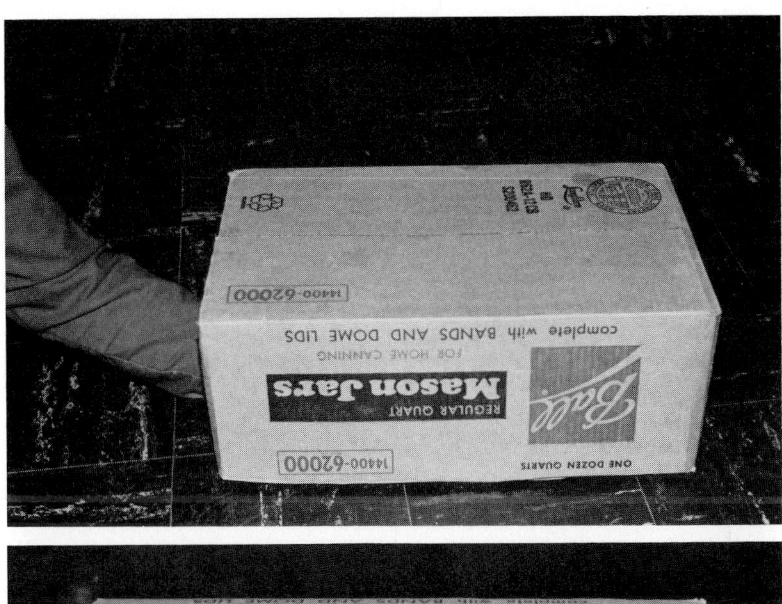

Fig. 4–1. Nested cardboard boxes muffle the sound produced by a .22-caliber starter pistol. Systematic desensitization is allowed as the nested boxes are uncovered one at a time.

adverse reaction when some of the muffling is removed. This therapeutic approach requires patient and consistent effort by the owner. In our sample case, Daisy's owner (Case 4–5) was willing to undertake an extensive treatment procedure.

CASE 4–5

History. Daisy, a spaniel cross, reacts emotionally to firecrackers. She has broken glass windows and severely injured herself. The cause of the behavior is unknown.

Diagnosis. Noise Phobia (firecrackers)

General Evaluation. The goal is to habituate the dog to loud noises through desensitization training sessions.

SPECIFIC INSTRUCTIONS

1. Do not give Daisy attention or comfort when she reacts with fear to loud noises. Try to act disgusted, displeased or indifferent.
2. Extraneous fear-producing noises will interfere with the desensitization training. Take her away during July fourth (e.g., camping in a state park). Alternatively, we can dispense a tranquilizer for July fourth.
3. Densensitization training sessions: fire starter pistol inside nested boxes where the sound is so soft she does not react. Reward her with a spoon of ice cream after each shot. Fire ten shots in a session once a day.
4. After 2 days remove one box. Remove a box thereafter every 2 days.
5. If she acts fearful during one of the sessions, return to a previously successful gradient level.
6. Progress check in 2 weeks.

Separation Anxiety

People and dogs are highly social animals that naturally experience some stress when isolated from companions. Separation anxiety is reduced in children when we tell them we will be back in a few hours, but we cannot communicate this to a dog. The longer we are gone, the more intense the separation stimulus becomes. An animal learns we will return only after it has experienced a large number of departures and returns by the owner. In this way the dog's emotional reactions to being left alone are habituated.

The use of increasingly longer separations is the best way of habituating a dog's separation anxiety reaction (see Chapter 22). The owners should expose the dog to gradually increasing periods of separation, and reward the dog with food immediately before leaving and upon returning. The food reward makes the separation a positive rather than a negative experience.

History. A diagnosis of separation anxiety is based primarily on the occurrence of several types of misbehavior which might include chewing furniture or woodwork, excessive vocalization, and defecating in the house (when the dog is otherwise well house-trained). There should also be some indication that the dog becomes emotionally upset when left alone and is not being destructive out of boredom or unruliness. Destructiveness in the house that is not related to anxiety or emotional distress is dealt with in the next section.

Therapeutic Approach. To habituate a dog to being left alone, the dog should be subjected to an initially brief series of separations that gradually become longer. Prior to each separation, the dog should be given a large bone to chew. The long-lasting food reward creates an emotional state that overrides the separation anxiety reaction, as long as the separation is not for too long. Procedures must be established for determining if the dog remained calm when left alone. As long as the separations are successful, the duration of each separation can be increased. The separations must be shortened if there is any sign of emotional stress or misbehavior.

When the owners must be away for an extended period of time, the dog should be left with friends or relatives. Confining the dog to a kennel without human contact may intensify the anxiety. This may mean that the owners might have to take the dog to work or leave it with friends until the therapeutic program is completed.

Separation anxiety became a problem for our sample case, Callie (Case 4–6), when her young master grew up and moved out of the family home. She tolerated being left alone when the woman of the house went

CASE 4–6

History. Callie is a 5-year-old spayed female Vizsla who will chew up almost anything when left alone for an hour or more in the afternoon or evening. The behavior started six months ago when the son of the family married and moved away. If left outside unexpectedly, Callie chews the shingles by the back door. What is surprising is that Callie's behavior is excellent when she is left each morning while the woman of the house is gone for a half a day of work.

Diagnosis. Separation Anxiety

General Evaluation. Callie's fear when left alone will be desensitized.

SPECIFIC INSTRUCTIONS

1. Do not leave Callie alone in the house, except during training sessions. During the day she must be taken to your friend or relative's home or with you to work.
2. Conduct separation training sessions 2 to 3 times a day for a "safe" period of time, usually 15 minutes. During the week, sessions will be in the evening.
3. Grab the usual things you take when you normally leave the house, turn out the lights, and in other ways, mimic a normal departure.
4. Leave Callie a knuckle bone or other treat when you leave.
5. Booby-trap the cushions on the couch and one to two cushions on the floor with loaded mousetraps.
6. Gradually increase the length of the training sessions, looking for clues that let you know if the period is short enough. A sample time progression of: 15 min, 15 min, 30 min, 45 min, 30 min, 1 hr, 30 min, 1 hr, 1½ hr, etc., is recommended. Keep the time variable.
7. Progress check in 2 weeks.

to work each morning, but other absences resulted in extensive destructiveness, whether Callie was inside or out of the house. During treatment the owner conducted specific training sessions and also extended her absences each morning for work.

FEAR REACTIONS TOWARD PEOPLE AND DOGS

A genetic predisposition leads some dogs to have an excessive fear of other dogs, which is expressed as extreme timidity. A lack of early social contact with other dogs can also cause dogs to be fearful of conspecifics.

Adverse experiences with certain groups of people can produce a fear of those types of people, such as men or women in uniform. In one instance, whenever the owner wore a motorcycle outfit, which included leather jacket, helmet and boots, her dog became extremely fearful, but recovered when she removed the outfit. Apparently the dog was specifically sensitized to people dressed in classical motorcycle attire. Fear of children frequently results from lack of early social experience with children. Objectionable urination, a sign of submissive behavior, sometimes indicates a fear of people. As with other phobias, both systematic desensitization and counterconditioning are used (see Chapter 22). In the case of the leather-jacket phobia, the woman desensitized the response, enlisting friends to assist in the training by wearing only the jacket or helmet and coaching her dog to approach, and rewarding the approach behavior when the fearful signs were absent. Eventually her friends progressed to wearing the full attire.

Fear of Other Dogs

By the time a client brings a fearful dog into the clinic, he can usually identify those situations that will elicit a fear response. Leading questions can then quickly clarify the particular circumstances in which the fear is involved.

History. When a puppy has been separated from contact with other dogs at the time when socialization is crucially important (see Chapter 21), the dog, as an adult, may develop a fear of other dogs, and act submissive around them. It appears that a lack of early social interaction deprives a dog of the ability to develop appropriate social responses.

Therapeutic Approach. There is no easy way to correct a dog's fear of other dogs. If the dog is forced to come into close contact with other dogs, the behavior will become worse. Achieving a gradual adaptation to those dogs that reside in the problem animal's immediate environment is the best one might expect to accomplish. The best therapeutic approach is to provide reinforcement in the form of affection and food treats when the dog moves toward other dogs. Since the owner needs the cooperation of the other dog owners in the area, solving this behavioral problem could be complicated. If the owner solicits the cooperation of other dog owners who can keep their dogs under control, the problem dog can be worked with, and the behavior perhaps improved. If the stimulus dogs are of a friendly nature, the likelihood for improvement is enhanced.

Fear of Certain People

When dogs are fearful of certain people, the practitioner can work with the client to define the fear stimuli as specifically as possible. This may involve verbally reviewing several episodes where the dog's fear has been expressed.

History. When treating a dog's fear reaction to people, it is important to learn about the boundaries of the stimulus configuration that create the problem and how these might be manipulated. Perhaps distance is important, or, alternatively, the appearance of the stimulus.

Therapeutic Approach. Two gradients that will affect the success of the desensitization of the dog's fear of people are the degree of similarity of the training stimulus to the fear-evoking stimulus, and the proximity of the training stimulus.

As in all fear reactions, forcing a dog to approach the fear-inducing stimulus is counterproductive. Thus, when the distance gradient is used, the dog is exposed to a fear-evoking person who is far enough away not to evoke a reaction. In training sessions the dog is repeatedly exposed to the fear-evoking person at a safe distance and immediately given a food reward. Affection and attention accelerate the desensitization process, particularly if the owners of the dog only give it attention during training sessions. This procedure lays the groundwork for later sessions when the dog receives attention and affection only from the person who previously evoked the fear reaction. Periodically, the dog should have its progress tested by having the owners withdraw completely while the test person comes within a safe distance and gives the dog an opportunity to approach him and receive some mild degree of petting and attention. At this point, the dog's own behavior becomes the guide as to how rapidly owners can move in training sessions. Once the process of approaching people gains momentum, the dog's natural inclination is toward social contact. Occasionally a dog is presented for consultation who has been exposed to only a few people. Cuddles (Case 4–7), for example, was raised by a dog breeder. While the owner typically had twenty dogs on her ranch, she had almost no visitors. For Cuddles, humans other than the owner were a novel stimulus.

As with other phobias, a dog's fear of people can sometimes be so extreme that it precludes any training session, as in the case of Shady (Case 4–8). A tranquilizer was useful in this situation to calm the dog so that systematic desensitization training sessions could be started.

CASE 4–7

History. Cuddles, a 2-year-old female Tibetan Terrier, is acutely fearful of all people except the owner, who breeds dogs. Cuddles has never been exposed to humans other than the owner on a regular basis, and as a puppy she had virtually no human contact. She gets along well with other dogs.

Diagnosis. Phobia to Strange People

General Evaluation. During treatment Cuddles will be given nonthreatening exposure to a variety of people. Training sessions will desensitize her fear reaction when people move closer to her.

SPECIFIC INSTRUCTIONS

1. Give Cuddles routine exposure to people at a distance. Take her daily to the pet store where you work, along with another dog as a social facilitator. Provide her a safe retreat where she can avoid people.
2. Initiate training sessions (10 trials/day) at work. Have the sales clerk move within five feet of Cuddles (or a distance where Cuddles is not afraid). Reward Cuddles with food. Reduce the distance by one foot each day, continuing to reward desirable behavior. Finally, the sales clerk may be able to gradually reach out toward Cuddles. Next, begin the series of training sessions again with a new person.
3. Avoid evoking a full-blown fear reaction, and stop all comforting of the fear. If it occurs, be indifferent.
4. Progress check in two weeks for further training.

CASE 4–8

History. Shady, a Siberian Husky, is extremely fearful of people. He is timid, except with familiar people. He will stand in the show ring, but it is obviously uncomfortable. At his worst, he urinates and defecates uncontrollably.

Diagnosis. Phobia to Strange People

General Evaluation. Shady will be tranquilized and then given desensitization sessions using counterconditioning.

SPECIFIC INSTRUCTIONS

1. Ignore Shady's fear response whenever he displays it.
2. Administer Valium, a 2 mg tablet 3 times daily. Double the dose twice if needed.
3. Create nonthreatening training sessions. Allow a friend to approach Shady until he almost manifests slight fear. The friend will then retreat, and approach again. Repeat for 10 trials. Present a food treat and praise him during each trial.
4. Return visit in 2 weeks.

Submissive Urination

While some dogs express their fear or anxiety by whimpering, cowering, or even barking, others urinate. This symptom of fear irritates and perplexes clients, who may not have identified it as fear-related and have no idea what to do about the problem. This problem is also covered in Chapter 8 in a discussion on elimination problems.

History. Timid and submissive behavior is normal for puppies in the company of adult dogs or people. In fact, a display of submissive behavior is typical of the species as a way of inhibiting aggressive approaches by other dogs. Urination upon being approached by either dogs or people is one aspect of submissive behavior. A vicious cycle can be started by owners of puppies who do not like the urination and scold their puppies for urinating. Since shouting and physical punishment are perceived as aggressive approaches by the puppy, it acts even more submissive to reduce the aggression and it urinates all the more.

Therapeutic Approach. To correct submissive urination, it is important to refrain from actions that make the dog more submissive. It is necessary to determine as exactly as possible what circumstances and which people evoke the behavior. For example, boisterous greetings may evoke the urination and quiet greetings may not. Perhaps greetings in the afternoon, but not the mornings, evoke urination. Once the least and most evocative situations are known, the next task is to plan a series of graded training sessions, starting with the safest situation and continuing to those that are more critical. The rewards should be given with some detachment, thus a food reward is usually more appropriate than affection. As soon as a training pattern is established so that the puppy receives food rewards during safe encounters and does not urinate, the trials can be phased gradually into situations more like those that evoke urination. For example, this might mean gradually changing the time of day when greetings are made, or gradually decreasing the proximity to the person who evokes the urinations.

Mild remote punishment may be effective in dealing with submissive urination. A water sprayer is the best device, but the dog must be sprayed such that it does not know that a person is causing the spraying. The technique is to have one person gently approach the dog while an accomplice gets ready to surreptitiously squirt the dog if it urinates submissively. The person approaching the dog should then turn and walk away. After a few sessions the dog may not urinate and, of course, its good behavior should then be rewarded. The initial sample case on Shabby (Case 2–1) added this type of apparently remote punishment for submissive urination. After the urinating dog was sprayed by an accomplice, the person holding the dog pretended to be very startled and suddenly dropped the dog to the floor.

Chapter 5

Activity, Barking, and Destructiveness Problems in Dogs

Every dog owner encounters problems with his pet, such as excessive barking, exuberance, or destructiveness. Usually these problems are just annoying and of minor consequence. At times, however, a dog's barking may become so objectionable that it is bothersome to neighbors, and the owners must consider getting rid of the dog. Household destruction may reach a point where several hundreds of dollars worth of furnishings are ruined. At these times professional advice may be sought.

The causes of misbehavior discussed in this chapter are: boredom, separation anxiety, excitement from interactions with neighboring dogs, and the simple rewards that come from engaging in misbehavior.

Frequently, these behavioral problems will arise simultaneously in dogs with a high level of reactivity (Hart & Hart, 1985). Generally, the therapeutic approach is similar for all of these misbehaviors. Because they often occur together and are treated similarly, they are included here within a single chapter.

EXCESSIVE ACTIVITY

Most people expect puppies to be very active. When a person has children, puppies are an excellent source of entertainment. In fact, a puppy is one of the best baby sitters a person can get. But when an adult dog is continually buzzing around, people find the behavior almost intolerable. Extreme activity is typical for some breeds of dogs, and the best we can do is try to enhance the problem dog's quiet periods and not increase the problem by reinforcing activity. In addition to breed predisposition, there is a form of hyperactivity called hyperkinesis which appears to be abnormal and may be amenable to drug treatment.

Diagnostic Problems

Of the three basic considerations for treatment, the first is to rule out a pathophysiological cause. This is unlikely, but it should be part of the diagnostic work-up. The second task is to determine whether one is dealing with puppylike activity or the innate tendencies of some breeds to be very active. The third task is to assess the degree to which the behavior seems amenable to conditioning procedures that would enhance calm behavior. The best measure of this is how often, and under what conditions, the dog is calm on its own. The applications of conditioning principles to enhance quiet behavior are discussed below.

Abnormal Hyperactivity (hyperkinesis). Dogs with this syndrome are so abnormally active in comparison to breed standards, that they exhibit truly abnormal behavior. They may also be uncontrollable and aggressive toward human handlers.

Some people have compared this type of hyperactivity to the hyperkinetic syndrome of young children. Central nervous system stimulants such as amphetamine drugs and methylphenidate (Ritalin, CIBA) have a suppressive effect on the hyperactivity of both dogs and children. Afflicted children appear to outgrow the disorder at the time of puberty, whereas the cases of hyperactivity reported in dogs have been reported only in adult animals; but this may or may not be an essential difference.

The first observations of hyperactivity in dogs were by Corson and coworkers who discovered that if such dogs were treated with an amphetamine, the hyperactivity was suppressed so that handlers could socialize with the dogs and work with them in obedience training (Corson, *et al.,* 1972). After the training, the drug treatment could be ended in some dogs with no return of symptoms. Corson also noted that amphetamines suppressed heart rate and respiration, as well as the general activity level of dogs.

The practitioner should take a conservative approach when dealing with this syndrome, and diagnose a dog as hyperactive only after more likely explanations, such as a lack of control or training by the owner, have been explored. If treatment with an amphetamine or methylphenidate is chosen as therapy, the practitioner should adjust the dosage according to the needs of each individual dog.

Excessive Activity Related to Breed Predispositions and Play. Sometimes people who are accustomed to the slow pace of an older dog feel that the behavior of their new puppy must be abnormal. Usually the owners of such hyperactive dogs do not realize how active a young retriever or boxer can be, and they believe the dog may be abnormal when the behavior is typical. One approach to problems of this type is to help the owners realize that their dog is normal, and that they may be enhancing the behavior by giving the dog attention when it is particularly active. Examples of such reinforcement are attempting to hit the dog, but missing or pushing it away when its activity is directed toward people.

Treatment of Excessive Activity

History. A determination should be made as to the type of excessive activity. Assuming that the problem is not physiological, attempts can be made to use conditioning approaches to diminish activity level.

Therapeutic Approach. This type of objectionable activity has been discussed by Voith (1980a), who suggests some straightforward remedies: When the dog's activity or play becomes objectionable, the owner should simply walk away and totally ignore the dog. Since young animals and dogs of certain breeds do require a certain amount of energy expenditure, the dog should have regular opportunities to engage in acceptable types of activity, such as jogging or playing.

Punishment can be used to stop ongoing activity that is objectionable if the punishment is delivered as soon as the activity begins and if it is employed every time the behavior occurs. Voith cautions that dog owners only occasionally punish their dogs, and even then, the punishment may be administered long after the dog has engaged in objectionable hyperactivity. Thus, the dog may "play the odds" and continue to be just as active as always.

If there are times during the day, or in certain places, where a dog is relatively quiet, it may be amenable to conditioning. For example, the dog may be quiet for a few minutes after it is told to "be quiet," or if a particular member of the family is petting it. If a dog which is too active is quiet in the examination room, this it is an indication that the quiet behavior may be shaped.

The principle of successive approximation can be used here to countercondition good (quiet) behavior. If, for example, the dog is quiet for two minutes when someone says "be quiet," this is used as a starting point. The dog is asked to be quiet when the owners know it is likely to be. Food treats are given to the dog when it has been quiet for at least 2 minutes. For several days the dog is asked to be quiet, and when it has been, it is consistently rewarded. Since affection and praise may excite the dog, it is probably best to stay with food reinforcement for this conditioning.

Over several days a longer duration of being quiet is required before a reward is given. Depending upon the success of the initial conditioning sessions, the goal of having the dog remain quiet for a half-hour or more can be approached over the next few weeks.

Remember that excessive activity probably reflects a breed or individual predisposition and, therefore, conditioning procedures can be expected, at best, to decrease the frequency of excessive activity but not to eliminate the problem. When a hyperactive dog exhibits a number of objectionable and destructive behaviors, as was the case with Blacky (Case 5–1), it may be necessary to interrupt these behaviors with remote punishment before initiating training sessions to condition good behavior.

EXCESSIVE BARKING

One of the advantages of having dogs is that they bark at stangers and warn us of intruders. This behavior is genetically acquired and is related to living in a pack, where there is an advantage to warning other

CASE 5–1

History. This hyperactive, castrated male Doberman Pinscher, Blacky, jumps on windows, rips screens, barks, and never sleeps when people are around. As a result of gobbling clothing, he has had a gastrotomy.

Diagnosis. Excessive Activity

General Evaluation. Remote punishment will be used to interrupt Blacky's specific objectionable activities. An exercise routine and obedience training can improve the general pattern of misbehavior.

SPECIFIC INSTRUCTIONS

1. Attend an obedience class with Blacky where you handle the dog.
2. Run every day with Blacky for the next two weeks. If this improves the problem, set up an increased exercise routine on a permanent basis. Be consistent with a daily routine.
3. Booby trap with mousetraps some things he is likely to chew or swallow. Hide the mousetraps well, and stay out of range so he doesn't suspect you doing it. As he catches on, vary the type of clothing.
4. Keep all unbooby trapped clothing away from Blacky.
5. Provide a variety of acceptable toys for Blacky to chew.
6. Return in 2 weeks, when Blacky is more controllable, to plan a conditioning regime.

members of the pack about intruding conspecifics. Dogs also treat people as conspecifics, and warn us of intruders, human or canine. Interestingly, dogs do not usually warn us of intruding birds, rabbits, or cats, presumably because such animals are not a threat to their territory. The tendency of dogs to treat people as they would other dogs is probably a function of the early socialization to people that a puppy receives from the exposure to a family or breeder.

Normal Barking Behavior

Barking is perfectly normal behavior for dogs, but it is a frequent subject of complaints. Aside from the biting dog, the barking dog is probably the greatest pet peeve of people who do not own dogs, and it creates many problems for dog owners. Barking behavior is an innately programmed response that is hard to eliminate with either punishment or conditioning procedures. In fact, it may be advisable to forget about eliminating barking and deal with the behavior by altering the situations that lead to barking. A dog that barks continuously in a backyard run while the rest of the family is away may not bark if left in the house. The owners may object if the dog is shedding, but it may be easier to deal with the shedding problem than to change the barking behavior.

Why do dogs bark? First, barking is a form of vocal communication. Dogs may also learn to bark if barking is paid off with rewards. If a dog barks long enough, for example, it may be allowed in the house.

Therapeutic Approaches

In selecting a therapeutic approach for barking, the first task is to determine if the behavior reflects a natural (inherited) tendency or is learned behavior which has been rewarded in the past. Work with the client to understand the present or past rewards available to the dog for barking. Learning how the barking grew into a problem is useful. Treatment of barking can involve extinction procedure, punishment, or counter-conditioning, depending on the nature of the problem.

Extinction. Extinction is a way of eliminating a behavioral problem that has been learned, but one must identify the reinforcing factors that initiated and maintain the behavior. The dog that barks upon hearing the command "speak," for example, may be rewarded with food. If the dog is never given food again when it barks after being told to "speak," it will eventually stop this type of barking.

In order to be extinguished, the behavior must occur again but never again be rewarded. Extinction might be used with a dog that barks to get into the house. When the owners do not allow the dog in the house until the barking becomes very frequent and loud, they shape the barking to a level that is objectionable. However, if the dog is never allowed in the house after a barking episode, but only when it has been quiet for five minutes, the barking will be extinguished.

Punishment. Some form of punishment may be the most expedient way to deal with barking, whether it is learned, spontaneous, or innate behavior. Interactive punishment, such as shouting at the dog or throwing something at it, is usually ineffective. If a person shouts at a dog or hits it, he arouses autonomic reactions, escape attempts, or submissive responses that interfere with the intended purpose of punishment. Remote punishment that immediately follows the specific act of barking and is not associated with the person who administers it is often the most effective way to stop barking.

Collars that deliver an aversive stimulus such as a shock or loud noise are an attempt at remote punishment. If a dog can be given an aversive stimulus each time it barks, then barking is likely to be eliminated. In this approach, the punishment itself is directly paired with barking.

The use of electrical shock collars has proven quite effective. At times, however, the shock collars will not operate at all. At other times they may be triggered by extraneous sounds or the barking of other dogs. If the collars are placed on the dog too loosely, the electrodes will not make contact and the dog is able to bark wihout receiving shocks. There are also instances of the electrodes burning the skin at the point of contact. Changes in resistance between the electrodes and the dog's neck depend on whether the dog was is wet or dry, and pose a considerable problem.

Recently, the electrical shock collar for barking has been replaced by a collar in which the dog's barking automatically activates a shrill sound.

These collars have not been on the market long enough to allow for an evaluation of their effectiveness. Remote control shock collars, triggered by a handheld console, are also available. This instrument allows the owner to observe the dog's reactions to the shock, and any untoward effects of excessive shock are immediately obvious. Also, there is no accidental triggering of the shock collar.

Dog owners can become creative with various forms of remote punishment. One frantic dog owner, who suffered through his dog's barking in a kennel late each night, devised a remote punishment system consisting of a platform which held a bucket of water directly over the kennel. Through a system of ropes and pulleys the owner could trip the bucket of water from his bedroom window, dousing the dog when it started to bark. Once it was dumped, the bucket automatically turned upright and could be refilled remotely by the owner turning on a garden hose. After just a few dousings, the dog's barking behavior was eliminated. A more practical type of remote punishment is a water hose with a power nozzle to spray the dog surreptitiously when it begins barking.

An important rule is that once the owner begins a remote punishment training program, every attempt should be made to punish all barking episodes from that time on. Periods of barking without punishment work against the entire punishment approach. Until the problem is solved it may be necessary for the dog to be taken to work or kept indoors, so that it is not allowed to engage in unpunished barking.

Barking is a misbehavior that commonly occurs with other misbehaviors, as in the case of Sass (Case 5–2). A practitioner must then decide whether to address a single misbehavior at a time, or to initiate a multipronged attack on problem behavior. With Sass, the owner was looking for an all-out approach to extinguish Sass's pervasive pattern of misbehavior.

Counterconditioning for Nonbarking Behavior. An alternative to the use of extinction and punishment to control barking is to attempt to employ counterconditioning to make the dog quiet through the use of rewards. This approach might be useful in instances where a dog barks persistently while the owners are not home. The process is time-consuming and one should consider advising the use of punishment first.

In counterconditioning we assume that there is a short period of time in which barking does not occur. This might be 10 minutes, for example. In training sessions the owner is instructed to leave the dog for periods ranging from 5 to no more than 10 minutes. If the barking has occurred only when the owners have left the house, then they should leave the house during training sessions. The dog's barking must be monitored by a person staying nearby while others leave, or by tape recorder. Upon return, if the dog has not barked, the owners should heap praise, af-

CASE 5–2

History. Aggressiveness, barking, and urine marking are among the array of problems exhibited by Sass, a castrated male Australian Terrier. Sass obeys general commands, but bites if food, tissues, or toys are taken from him. He barks if stared at, or if the owner is leaving unexpectedly.

Diagnosis. Dominance-Related Aggression, Barking, Urine Marking

General Evaluation. Both pharmaceutical and behavioral treatments will be used. The aggression and urine marking will be treated with a progestin and interactive punishment. We will use remote punishment to treat the barking and the urine marking.

SPECIFIC INSTRUCTIONS

1. The medroxyprogesterone (150 mg) injection may decrease the aggressive dominance and urine marking.
2. For aggression, directly punish misbehavior by jerking him off the ground by his choke chain lead. Leave a short lead attached to the choke chain for easy access to Sass.
3. Give Sass easy commands to obey and reward him for good behavior with food treats and affection.
4. For barking, use remote punishment. Ambush Sass with a water sprayer so that he doesn't know you are doing it. Reward Sass if he is quiet for 30 minutes at a time.
5. For urine marking, punish immediately after the occurrence. As remote punishment, booby trap his typical spots with mousetraps.

fection and food treats on the dog. Affection withdrawal during the day may enhance the value of affection as a counterconditioning reward. After several days of these initial practice sessions, the owner should notice that the dog seems to anticipate the owner's return and to expect a food reward.

In subsequent training sessions, the dog is left in the situation for gradually increasing periods of time, ranging up to an hour or more. It will still be necessary to monitor the dog's barking by getting the help of a neighbor or other volunteer. Once the counterconditioning program described above is initiated, the dog should never be left in a nontraining situation where it is likely to bark. This may require taking the dog to work or leaving it with neighbors. Assuming that the increments of time the dog is left alone are increased gradually, the dog learns that quiet behavior begets rewards. It may be happy to remain quiet as long as the owners continue to reward nonbarking behavior.

If barking is much too easily triggered for the owners to pursue the counterconditioning program, it may be necessary to start with the dog continuously tranquilized for the first week or so. A phenothiazine tranquilizer is the drug of choice (see Chapter 24).

DESTRUCTIVE BEHAVIOR

Digging holes, chewing on furniture, and clawing into the side of a house are common complaints. When these behaviors are displayed by puppies we tend to excuse it as normal puppy behavior and cope by keeping things out of reach until the puppy grows up. In an adult dog, destructive behavior is a different matter. Adult dogs are large enough to do real damage and, if uncorrected, the behavior is likely to persist. When the owners leave home or are not around the dog, the destructive behavior is most likely to occur. Interestingly, the easiest form of destructive behavior to treat is that which happens predictably each day. An effective punishment or conditioning program can then be designed. If the destructiveness only occurs sporadically and is impossible to predict, it can be very difficult to resolve.

Diagnostic Approaches

Treatment of destructive behavior depends on an analysis of the emotional state during acts of destructiveness, analysis of when the destructiveness occurs, and the type of destructiveness.

Analysis of Emotional State. A dog left alone may be tearing into things as an outlet for boredom, because the exercise is intrinsically rewarding. Alternatively, there may be some tangible rewards, such as food scraps for a dog that gets into garbage cans. These motivating factors would not appear to involve anxiety or fear. If the dog appears very anxious and upset when left alone, its behavior probably represents a form of separation anxiety, and the destructiveness must be dealt with as secondary to the anxiety. Separation anxiety and its related destructiveness are dealt with in Chapter 4.

When there is no apparent anxiety involved, either punishment or conditioning of desirable behavior is a feasible therapeutic approach, or a combination of both may be used. The dog can be physically restricted on a leash, or left in an enclosure, when it must be alone for extended periods until the problem behavior is resolved.

The cause of the behavior, perhaps boredom or a need for activity or general discipline, should also be explored. A regimen of exercise, obedience training, and daily sessions where the owners interact with their dog may aid in therapy and prevent recurrence once the behavior is controlled. However, exercise or attention sessions should be scheduled only if they are a daily routine, and not just on weekends and the occasional week day. Inconsistent interactions may evoke undesirable destructive behavior as the dog becomes more frustrated in its attempt to find an opportunity for exercise.

One useful clinical sign in diagnosing anxiety is if the dog engages in several types of misbehavior such as excessive barking and house soiling in addition to destructiveness. When dealing with anxiety, the emotional state must be habituated or extinguished by systematic desensitization. This means gradually getting the dog used to being left alone for longer and longer periods of time. Restraining the dog in the home when it must be left alone for long periods is a useful practice in dealing with destructiveness related to general unruliness, but when the destruction is related to anxiety, the isolation makes the anxiety worse. Therefore, the dog may have to be taken to visit relatives, friends, or even to work when it is not left alone for reasons other than departure training sessions. Keeping a dog constantly tranquilized for a few weeks during training sessions is also a possibility in dealing with separation anxiety.

Analysis of When Destructiveness Occurs and Type of Destructiveness. The simplest case is when a dog misbehaves every time it is left alone for a short while. Since the behavior is predictable, punishment and positive conditioning sessions can be prescribed. When the destruction is sporadic, some background observations are necessary. The owner may find, for example, that departures in the morning, particularly if the dog was taken on a walk prior to departure, are much more likely to lead to misbehavior. There may be some other factors involved, such as the presence or absence of other dogs. Sometimes loud noises, such as airplanes taking off, or the occurrence of sonic booms, stimulate a dog. If this is the case, the anxiety produced by the loud stimulus will have to be addressed in eliminating the destructive behavior.

Is the dog's destructiveness limited to one area, such as repeatedly digging a hole in the same place, tipping over the same garbage can, chewing on the same area of fencing? If this is so, remote punishment may be a useful therapeutic approach. If the misbehavior varies so that there is a different target each time, punishment is inappropriate. Remote punishment is useful only when one or two places are target areas. When the destructiveness is directed at a different area each time, this is an indication that emotional anxiety may underlie the behavior.

Therapeutic Approaches

Two completely different approaches are used to treat destructive behavior: punishment of inappropriate behavior or reward of good behavior.

Punishment. This is usually the most direct approach and may be used when there are no strong emotional reactions involved. Punishment is particularly indicated when the problem is quite specific. For example, a dog may dig repeatedly in one area of the yard or chew in one specific place on a fence. By placing loaded mousetraps on the spot in the yard which is repeatedly dug out, or attaching a livestock fence charge (hot wire) to the object which is repeatedly chewed, one is utilizing remote punishment. The punishment is delivered immediately after the misbehavior and the punishment is directly related to the target area. The snap of mousetraps, or the sting of a hot wire, is paired with the dog's own behavioral acts and not with a person. The punishment is immediate. This type of punishment is much more effective than hitting the dog some time after the occurrence of misbehavior. Even when hitting with the hand or a newspaper occurs immediately after the misbehavior, the connection between the behavior and the punishment is probably obscured.

If one has the opportunity to spy on a dog while it is being destructive, there are a variety of remote punishment tricks at one's disposal. For example, one might tie a bunch of tin cans together and string them up above the area where the misbehavior occurs. The tin cans could then be released at the onset of misbehavior.

Remote punishment is only useful when the misbehavior is directed toward a specific object. Obviously it is not practical if the destructiveness is constantly being directed at new objects. For this reason it is important to determine the location of misbehavior before deciding whether to recommend a punishment.

Punishment is not advisable for destructive behavior that is a result of fear or anxiety. If the animal already has an emotional reaction, adding a painful punishing stimulus to the situation may only worsen the behavior.

Counterconditioning. The reward approach is useful when a dog directs its destructive behavior toward a variety of objects or locations. Punishment is not feasible and we must think of ways to reward good behavior instead. This could be done in more than one way. For example, if a dog is digging holes in a number of places around the yard, it might be possible to condition the dog to dig where you want it to. Make a sandy digging spot in the corner of the yard and hide a bone just below the surface. Eventually the dog finds the bone by digging in the area. Next time bury the bone deeper and at subsequent times bury it deeper and deeper. Bury two or three bones and spread them out. Next, start hiding bones only sporadically to reinforce digging behavior on an intermittent basis. Meanwhile if the dog digs a hole in a nonacceptable place, punishment with mousetraps might be tried. By working with punishment of the occasional digging and intermittent reward of acceptable digging, one may successfully train a dog to dig only in the one allowable corner of the yard.

For household destructiveness and for digging problems one can also attempt to condition a dog to display good behavior by utilizing a training program outlined by Tuber, *et al.* (1974) and further amplified by Voith (1975a; 1975b). The procedure is to design a gradient of increasing duration when the dog is left alone. Begin with leaving the dog alone for a period of time so short that it is highly unlikely to be destructive. The departure durations are then gradually increased as long as the dog continues to be good. It is fairly typical for a dog to be good if left alone for ten minutes but to be destructive if left alone for two hours. The

CASE 5–3

History. This female Golder Retriever, Posy, chews everything in sight when the owners are gone. She shows no signs of anxiety or emotional upset when left alone.

Diagnosis. Destructive Behavior

General Evaluation. Treatment includes remote punishment for misbehavior and rewards for good behavior, supplemented with exercise and counterconditioning to reduce the destructiveness.

SPECIFIC INSTRUCTIONS

1. Use routine exercise to decrease her general activity. Be consistent in your daily schedule.
2. Administer remote punishment. Booby trap the areas she destroys with mousetraps. Set them daily so you know when she is good. If she chews a few times without punishment, the action is reinforced. Or booby trap as above using cayenne pepper. Or ambush her with a garden hose or spray gun. Do not scold her when you do this, as you don't want her to associate the punishment with you.
3. Countercondition Posy when you leave. Give her knuckle bones so she has good feelings and is distracted.
4. Give praise when Posy has been good. If when you return she has not chewed destructively, praise her profusely. If she has, withdraw attention for several hours.

goal is for the dog to be reliably well behaved for as long as eight hours. By gradually moving along the gradient, the duration of time the dog can be left alone is increased.

Once the baseline of planned departures is determined, a schedule of training sessions should be planned, starting with departures which are safe. The initial departures are designed to be short enough to preclude any destructiveness by the dog. Good behavior of the dog when it is left alone is rewarded by food treats and affection. Attention withdrawal for a time preceding the departure training sessions may enhance the effectiveness of the affection. If destructiveness does occur the owner should be cold to the dog upon return, isolate it for an hour or so, and the next day conduct a training session of shorter duration. It has been suggested that the dog be provided a new cue during training sessions, to which acceptable behavior can be associated. Tuning the radio to a program not normally listened to is appropriate. Between training sessions the dog must be physically prevented from being destructive whenever it is necessary to leave the dog alone for periods longer than those of the training sessions.

Planned departure training sessions can be employed if dogs are digging holes or destroying household items. It is best used for problems where the destructive behavior occurs almost every time the dog is left alone. If the destructiveness is infrequent, then it is probably not feasible to use this approach. A combination of remote punishment and rewards for good behavior was also used in our sample case with Posy (Case 5–3), whose misbehavior was chewing.

Chapter 6

Attention-Getting Behavior in Dogs

This type of behavioral problem can usually be handled with dramatic success. What's more, if you resolve the problem, your client will think you are a diagnostic genius, especially if someone else has wrestled with the problem unsuccessfully. Few behavioral problems are more fun to solve than attention-getting behavior.

What are attention-getting problem behaviors? They can range from vomiting, diarrhea, and anorexia, to snapping in the air, but the one thing they have in common is that performance of the behavior results in a pay-off, usually in terms of extra attention.

The initiation and maintenance of the behavior depends on the dog receiving reinforcements of attention, affection, or social contact. Many times a behavior such as acting lame or coughing has genuine medical cause to begin with, but because the behavior results in more attention than usual, the dog retains the behavior as a "gimmick" to sustain the extra attention. One of the more dramatic attempts for attention is to feign lameness. A dog that has injured a foot often receives affection and sympathy while it is injured. Being lame pays off handsomely in terms of more attention and affection, and the dog learns the value of pretending to be lame after the injury has healed.

Many times these medical signs could have been prevented. As Tanzer and Lyons (1977) mention, this is a good reason not to overlove a pet in the hospital when it is recovering from surgery, because it may continue to act out the sickness at home to get additional attention. Sometimes a dog puts on an attention-getting act in the presence of one member of the family, but then drops the act in the presence of other members.

DIAGNOSTIC APPROACH

Since animals have the ability to learn an infinite variety of behaviors, attention-getting acts almost defy categorization. Many weird acts such as chasing shadows, light beams, or imaginary flies are likely to be attention-getting, and are almost classic, as in the sample fly-snapping by Tommy (Case 2–2, pg. 24). A variety of would-be medical problems, including lameness, paralysis, muscle twitching, diarrhea, vomiting, ear problems, and conjunctivitis, have been documented as attention-getting. Anorexia can even be attention-getting, since the dog then receives pleading and hand-feeding as the owner tries to get the dog to eat. In some cases the behavior may be quite evident in the examination room (Figs. 6–1 and 6–2). At other times the examination room setting may inhibit the behavior.

Differential Diagnosis

A careful diagnostic workup for the possibility of a medical cause for attention-getting behavior is usually indicated. The diagnosis is often made by the process of exclusion after conducting an extensive series of clinical tests.

The tip-off is usually that the clinical signs or behavior are not consistent with other medical signs. The lameness persists in spite of every indication that the leg is normal. A postural twitch exists in spite of perfectly normal neurological examination findings. At this point you

Fig. 6–1. Attention-getting behavior. This dog would chase and bark at shadows or moving spots of light.

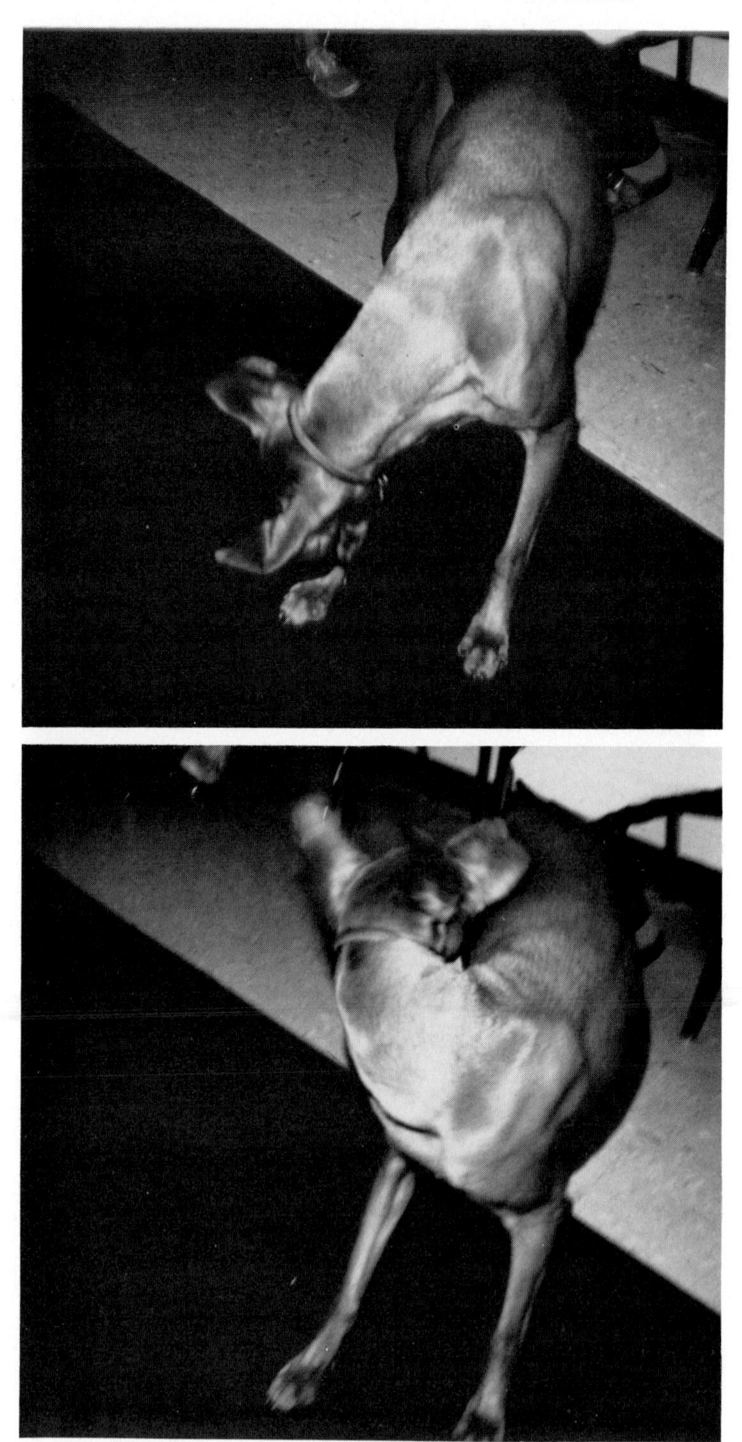

Fig. 6–2. Attention-getting behavior. This dog's act consisted of making movements back and forth between the postures seen in the top and bottom photographs.

have the choice of pursuing further clinical tests or exploring the possibility of an attention-getting cause early in the medical workup.

When considering an attention-getting diagnosis, obtain a medical history with some questions about what the owner does when the animal displays the behavior in question. Giving comfort and affection are obvious. However, by coughing a dog may induce the owner to allow it to stay inside rather than being put outdoors as the owners might prefer. Shouting can be rewarding since it is better than being ignored. About the only thing an owner can do that could not be construed as a reward is to completely ignore the dog.

Predisposing Factors

Attention-getting behavior is found in the typical household situation in which the dog is already heavily indulged with love and petting. While it may seem illogical, dogs that are getting the most attention are those that will go to some effort to gain even more. It is probably impossible to satiate a dog with too much attention. Dogs that receive little attention from their owners have the least probability of acquiring this behavioral problem.

Attention-getting behavior frequently occurs when there is more than one dog in the family. There is invariably some competition between dogs for attention from the human side of the family. If one of the dogs catches on to a gimmick that results in it getting more attention than the competition, it will stay with the behavior as long as there is a payoff.

The diagnosis of attention-getting behavior requires a bit of detective work to determine if the dog rarely displays the behavior when no one is around to reward it. One approach to test this is to hospitalize the dog. Some dogs that display their act only in front of their owners will act perfectly normal in the hospital. If the act continues in the hospital, have hospital personnel sneak up on the dog and watch it when it does not know it is being watched. A large, wall-mounted mirror might be used. If the dog displays the behavior when someone is obviously watching it, but not when it feels it is alone, this is conclusive evidence for the existence of an attention-getting motivation. A behavioral pattern that is secondary to a disease, or that is genetically based, will be displayed whether a person is around or not.

Generally, the tests of whether the behavior is attention-getting or not should be conducted at the owner's home. For such a test, the owners should prepare to leave as they normally would, including driving away in the car. At least one person quietly sneaks back while a partner drives away. The dog may have to be kept in one room so it can be easily spied on. If the owners can see the dog behaving normally when the dog does not know they are around, the diagnosis is pretty clear. This diagnostic approach is not as easily performed with some medical problems, such as diarrhea, that require more time for the problem to be analyzed. Boarding the dog with a friend may be the best approach.

An occasional display of the act when the dog is alone does not rule out an attention-getting motivation. A distribution of 95% when people are around and 5% when they are not would still be consistent with an attention-getting diagnosis.

THERAPEUTIC APPROACHES

There are three elements involved in treating the problem. One is to extinguish the attention-getting behavior by removing all pay-offs, the second is to reinforce desirable behavior, and a third is to remotely punish the attention-getting behavior.

Before instituting these therapeutic measures, recognize that owners are often reluctant to believe the problem is behavioral and not pathophysiological. Therefore, it is usually necessary to fully explain what has been going on. If the owners have been able to spy on the dog and can see for themselves that the behavior does not occur when they are not around, this is enough to convince them to follow the instructions. If you have hospitalized the dog, and noticed that the ailment was not displayed for an entire week at the hospital, this information may also convince the owners.

The first measure, which is the most important element in the therapeutic approach, is to extinguish the behavior. This can be accomplished by absolutely ignoring the dog when attention-getting behavior occurs. If the behavior is going on most of the time, the owners will have to persistently ignore the dog. When good behavior occurs, such as sitting and being still, and even sleeping, the owners should give affection, attention and even food treats to the dog.

Ignoring a favored family pet is very difficult for some people to do. An approach that frequently works with getting reluctant owners to accept advice is to suggest they try an experiment for one or two weeks. There will usually be substantial progress in one week, and owners are then willing to continue to follow your advice.

Ignoring a dog when it displays the problem behavior, but rewarding it when it is good, is just the opposite of what was occurring before. Previously, when the dog was engaging in the problem behavior it got attention, and when it was lying still, it was left alone.

If there is one thing dogs crave, it is affection and attention. As time goes on, a contrast is created between behavior that begets attention, and behavior that begets indifference and neglect. The more the contrast can be emphasized, the more readily the behavior will be altered. If the behavior is displayed only periodically, a good ploy is for the owners to leave the dog alone and even leave the house immediately on onset of the behavior. This really gets the message across to the dog.

Sometimes attention-getting behavior gains its power because the owner is concerned that the dog has a severe medical problem. The client in these cases may be dumbfounded or feel betrayed to learn that the dog has successfully played such a trick. The client may then feel foolish, and this actually makes it easier for the client to begin ignoring the behavior. For example, when Mrs. Simons learned that Contessa's behavior (Case 6–1) was an act, she automatically lost her great concern about the behavior and stopped attending to it.

Is punishment useful? Interactive punishment, such as hitting the dog or yelling at it, is not useful because this can be attention which simply maintains the behavior. Remote punishment, that which is disconnected from the owners, can be a useful adjuvant to the extinction therapy. This might involve the surreptitious use of a water sprayer as a mild form of punishment when the behavior starts. This works best if two people are in the room; one can usually squirt a dog without being seen. This not only gives the owners a way to vent their irritation when the objectionable behavior starts, but if the undesirable behavior is stopped for a few minutes, it allows the owners to reward good behavior. Whether or not remote punishment is used along with the process of extinction,

CASE 6–1

History. Mrs. Simons is exasperated with Contessa, a 3-year-old Golden Retriever who spends much of her time staring at the floor (as if pursuing a tiny insect). During these spells she secretes a profuse, watery nasal discharge that soaks her neck and forelegs, and she is nonresponsive to the owner. Since she may have consumed a large quantity of weed-killer around the time of the symptom onset, toxicity is a possible cause of the behavior.

Diagnosis. (1) Possible Attention-Getting Behavior; (2) Toxicity or Other Physical (Medical) Cause.

General Evaluation. Determine whether Contessa engages in nose rubbing and secretes a nasal discharge when she believes you are gone. If not, she is manifesting attention-getting behavior. If so, she may be suffering from toxicity or another medical problem.

SPECIFIC INSTRUCTIONS

1. Spy on Contessa while she believes you are gone. When you arrive home, greet her as usual, then leave promptly after she starts the behavior, but before she is wet. Take your keys, leave, and drive away so that she "believes" she is truly alone. Then sneak home and spy to see if nose rubbing and the nasal discharge occur.
2. Prepare for spy sessions, adjusting a drape or a stool or whatever, as needed, so that you can see her without being detected.
3. Spy 2 to 3 times, or until you know fully whether she engages in the unacceptable behaviors when you are gone.
4. Return for a progress check in one week. If Contessa is engaging in attention-getting behavior, we will formulate instructions for leaving when she engages in the behavior, remotely punishing it, and rewarding good behavior.

the object is to get the dog to be good more frequently or for longer durations of time before it is rewarded. In other words, the owners must be more progressively demanding of good behavior. Of course problems will vary. Some behavioral acts may go away overnight. Medical problems may take a week or two to clear up.

The behavior of a dog avoiding food or acting as if its appetite is lost requires special consideration. The dog is shunning one reinforcement, food, to obtain another reinforcement, attention.

Acting normal, that is eating, when the owners are not around would not be expected because the owner would know the dog was eating and would stop pleading. Therefore, the dog cannot eat much at all to make anorexia payoff. The diagnosis is based on how much attention is lavished on the dog to get it to eat, and on the absence of sickness to explain the behavior. A tip-off might be if the dog eats when kept by someone who pays little attention to it, but does not eat when cared for by the owner.

The treatment of ignoring the behavior must be balanced against the danger of the animal actually starving from undernutrition. We all know, for example, the extreme weight loss that can occur in humans in the psychologically based anorexia nervosa.

Therapy begins when the pleading is stopped altogether, as in the case of Andy (Case 6–2). The dog is offered a small dish of food and no fanfare is made. When behavior is in the direction of eating, such as just smelling or tasting the food, the dog is petted and comforted. When it turns away from the food, the owners should turn away from the dog. The idea is to reward any behavior related to eating with attention. If the dog does not smell the food but only looks at it, then this behavior may have to be rewarded. Once the dog smells, tastes, or eats just a little

CASE 6–2

History. Avoiding food is the problem that brings Andy, a male Golden Retriever, and his owners in for treatment. Since Mrs. Marsh recently started working, Andy is alone most of the time and gets much less attention.

Diagnosis. Attention-Getting Behavior: Anorexia

General Evaluation. Andy gains attention by eating less. You are to provide attention for good eating behavior and remove the reward for not eating.

SPECIFIC INSTRUCTIONS

1. Schedule some routine attention sessions for Andy. Play or exercise with him 30 minutes each day. Be consistent with the schedule.
2. When Andy does not eat, ignore him. When he shows interest in food and licks it, give him affection and attention. As times goes on be more demanding of the amount of eating before giving him attention and affection.
3. If these steps do not reinforce normal eating behavior, we'll consider administering a progestin to stimulate his appetite.

food, the criterion for rewarding can be made more stringent, such as eating a larger amount. The success of the program requires that attention throughout the day be withdrawn so that the dog only gets attention for good behavior related to eating. If it is necessary to force-feed a dog, this could be done by a nonfamily member in a detached manner.

Chapter 7

Roaming and Escaping in Dogs

Most dogs are forced to live in fenced back yards, kennels, or houses. Few people live in rural settings where they can allow their dogs to run free. Occasionally, dogs with access to wide-open spaces take advantage of this freedom and repeatedly run away for several hours or days. Correction of this problem requires some idea of what territory and home range mean to the dog.

REASONS FOR ROAMING AND ESCAPING

To understand the roaming behavior of dogs, it's useful to consider the wolf, the wild ancester of the dog. A typical pack of wolves has a home hunting range of several square miles that usually includes several den areas. Single wolves that are pack members may leave the pack for a few days. These adventures may arise from sexual interests, or from more mundane interests, such as hunger or exploration.

Dogs are similar to wolves in that they identify primarily with a pack and secondarily with a home. The household family is, of course, the pack. A dog will adapt to its human pack moving from city to city during a vacation, just as a pack of wolves will adapt to their den being moved from time to time.

The wolflike form of wandering does not directly correspond to the situation of a dog living in a home or even on a farm. A dog's home range may be the size of a small town. Although we might believe it to be abnormal for a dog to roam miles from home, the dog may find this quite natural. The dog has not forsaken its packlike bonds with the family for wanderlust.

Roaming has rewards. One is the powerful attraction of an appealing mate. Roaming for this reason is more likely in males than females. Food is a second reward. Meat or another interesting food treat may be as effective as sex in rewarding roaming. Or a dog may be drawn to social activity. When the household is empty, a playground with school children can be too much to resist. For similar reasons, a dog may also be attracted to the company of other dogs living some distance away.

Another type of problem, often associated with roaming, is escaping from a pen or kennel. If, as often happens, the dog manages to work its way out of the enclosure and simply hangs around the house, we are dealing with a problem basically different than roaming. If the dog both escapes and roams, we are dealing with a combination of the two problems.

Many dogs are kept in small fenced areas or backyard kennels, because their owners dislike having dogs in the house, or because they have both housedogs and outdoor dogs, or simply for convenience. While this arrangement is obviously not as much to a dog's liking as a more open habitat, most dogs accept kennel or small-yard life quite well. The introduction of the veterinarian to this problem may arise in having to treat wounds caused by the dog climbing over, under, or through the enclosure.

Most dogs seem to acquire a sense of territorial identification with the area in which they are enclosed, and they adapt to a routine of intermittently interacting with people on a scheduled basis. For example, they adjust to a routine of going for a walk or run each morning. As creatures of habit, dogs accept a consistent exercise routine, and they will usually not vocalize excessively or attempt to climb out of their enclosures at other times.

However, confined dogs are too often subjected to erratic routines. A dog that is never let out on weekdays may be taken out on runs on weekends and allowed to play with people in the house. The dog cannot discriminate weekends from weekdays, and thus it cannot settle into a routine. Since the dog lacks a routine and is unable to predict when it will be able to get out, it may repeatedly escape or attempt to escape.

As a dog yowls, digs, and chews at a fence, and even jumps and throws itself at the fence, the owner becomes even more inconsistent in his behavior toward the dog. The owner may severely punish the dog one time, but the next time unintentionally reward the dog by taking it inside the house until the kennel is repaired.

When dogs attempt to escape from an enclosure and repeatedly fail, they soon give up (extinction) because the effort is unrewarded. If the enclosure is rickety and a vigorous escape attempt pays off, further attempts at escaping are going to be all the more intense (shaping) (see Chapter 22). Minimal patching on such an enclosure will probably fail to prevent the dog from escaping again. The dog will try a little harder the next time, since it was only the most vigorous attempts at escaping that were reinforced before.

If the enclosure is made absolutely secure after two or three escapes by the dog, it will make many attempts to escape, but the behavior will eventually be extinguished. Extinction of escape behavior will be more rapid if the enclosure is made absolutely secure immediately after the first escape rather than after several escapes.

Therapeutic Approaches For Roaming

If a dog roams, the first requirement is to determine why the dog is leaving. Is it because everyone in the family is gone during the day so that the only social contact is several blocks away, or that someone else is feeding the dog? The reasons for roaming affect our choice of therapeutic approaches. If one can remove the source of the attraction, solving the problem will be greatly simplified. Other approaches range from castration to behavioral conditioning procedures.

Gonadectomy. Unless there is an obvious, nonsexual attraction for a dog's roaming, such as a playground or food, one should presume sexual motives for any roaming gonadally intact male. Roaming is one of the indications for castration, and castration is quite effective in reducing this behavior. In a survey of the effectiveness of castration for various types of problem behavior, castration was found effective in resolving roaming problems in 90% of the dogs surveyed (Hopkins, *et al.*, 1976). This rate of effectiveness was considerably higher than the 50 to 60% cure-rate for castration on intermale aggressiveness, urine marking, or mounting.

Aside from castration there are two other remedies for roaming, punishment of roaming, and reinforcement of nonroaming behavior.

Punishment. We all feel ambivalent when we catch the wandering dog and must decide whether we should punish it. Punishing the dog may cause it to escape from us when we go after it the next time. Yet, to pet the dog and welcome it into the car seems like rewarding it for roaming. It is probably safest to be indifferent when catching the dog and bringing it home, since hitting the dog is probably counterproductive.

If the dog goes to the same location repeatedly, there is a type of punishment that may be effective, but it requires the cooperation of others. If the people in the neighborhood where the dog roams can be persuaded to throw small rocks or use a blast from the garden hose on the dog, this is ideal punishment because the place to which the dog goes is punishing. Hopefully, the dog will avoid the locations that are punishing, and choose to stay home.

Reinforcing Good Behavior. Food can be used to reward a dog for staying home. A good reward is some favored food, like bits of chicken, pieces of cheese, or even steak. If the dog typically runs away during the day, it can be kept particularly hungry during the day by initially depriving it of food for 24 to 48 hours and then offering a distribution of 75% of its diet intermittently throughout the day. After this we can schedule food rewards so there is always a chance the dog may get a food treat any time of the day. For example, Buffy frequently ran off to run with other dogs, to visit a female in heat, or simply for a long trek across a neighboring farm. Since he had been castrated earlier, this was not an available solution.

Buffy loved fried hamburger, so the owner was instructed to offer Buffy a small snack of hamburger every half-hour for the first 4 days of treatment. Soon Buffy learned that he was getting super treats by just staying home. On the fifth day the client was told to vary the time between treats from 30 minutes up to 1 hour. As long as the training was successful, the owner was to gradually increase the time between food treats to two hours, and to even visit the dog sometimes without a treat. In the final stage of treatment Buffy was to be given treats more sporadically, but only a few a day, just so the dog would continue to stay home.

Buffy responded well to the treatment. He remained home almost all of the time with only three exceptions during the initial days of treatment. The owners continued the sporadic food treats and Buffy became a contented housedog. A similar treatment regime was effective for the sample case, Jack (Case 7–1), who lived on an open ranch and frequently ran away.

To summarize, dogs quite naturally explore the environment around their homes, and to them this probably seems quite appropriate. We can decrease the likelihood of them roaming by (1) gonadectomy, (2) reducing the appeal of the places they visit, and (3) making it appealing for them to stay home all day. As a final resort, some dog owners may decide to confine their dogs to a pen or kennel. If this choice is made, it is wise to review with clients the behavioral concerns one should have in keeping dogs closely confined.

CASE 7–1

History. Jack, a Labrador Retriever, lived on an open ranch. He frequently ran away to visit other dogs for a long trek in the woods. Jack was castrated a year ago with no effect on the behavior.

Diagnosis. Roaming

General Evaluation. Jack will be rewarded intermittently for staying close to home.

SPECIFIC INSTRUCTIONS

1. Withhold food for 24 hours. Provide food rewards for Jack staying close to home by giving him small pieces of chicken on a scheduled basis.

Days 1 and 2: Every 15 minutes.
Days 3 and 4: Every 15 to 45 minutes, on a variable schedule.
Day 5 and on: Continue increasing the intervals between treats, and keep the intervals variable.

2. During treatment, feed Jack his meals only in the evenings when he tends not to roam. This will keep him looking for daytime treats.

Therapeutic Approaches for Escaping

Let us again review how we train canines to escape from their kennels. If the dog owner does not make the enclosure escape-proof despite numerous escapes, a more extreme form of escape behavior will be shaped. For example, the owner may do a slightly better patch job each time, as when chewed wood is repeatedly replaced and reinforced by chicken wire. Although the chicken wire is more difficult for the dog to chew, the dog has been conditioned to persist in escaping, and eventually it gets through the wire. If the chicken wire is then replaced by hardware cloth, more effort is required, and the dog must take more wear and tear on its jaws to get out. If the dog escapes again, the owner may then put up chain-link fencing. Now, however, the dog's moderate escape attempts have been extinguished and only fierce biting and pulling behavior have been rewarded. The dog's escape attempts have been shaped to the extreme limits of its physical abilities. Normally dogs will not cut through chain-link fencing, but they can perform some unbelievable tasks of strength and endurance if their behavior is shaped.

The case of Wolfgang serves as an example of an escaping dog. Wolfgang threw himself at his chain-link enclosure. He would rip and tear at it with his teeth until he was out. His feet were torn and infected from jumping against the fence, and one leg was lame. His body was wounded and infected from squeezing through tiny cracks in the fence. Wolfgang escaped regularly, and then quickly returned home after briefly visiting a female dog who lived across the street. Wolfgang's escape behavior had been shaped by the owner, who each time made it a little more difficult to escape, but not impossible. Wolfgang also learned that once the enclosure had a new hole in it the owner would allow him in the house during the repair period. This reward of staying in the house added motivation for Wolfgang to escape.

Reinforcing Good Behavior. First, the most important detail is to maintain a routine in removing the dog from the enclosure. Secondly, once a dog begins escaping, it is necessary to make the enclosure totally escape-proof to extinguish the escape behavior. Ideally, the enclosure is made secure immediately after the first escape so as not to reward more intense escape behavior and increase the animal's resistance to extinction behavior. The more frequent the dog's escapes, the longer it will take to extinguish the behavior. The dog owner may consider the use of remote punishment if the escape behavior is directed toward just one or two places in the enclosure.

Punishment. Most people use punishment of some kind when their dogs get out. Most punishment, such as hitting or yelling, is interactive and does little to alter escape behavior. Another type of punishment, remote punishment, may be of some value if the dog's escape attempts are limited to one or two specific areas of the enclosure. Setting mousetraps around the escape area can produce a snap on a paw and a loud noise that may frighten the dog away from the area. A person can also use a livestock "hot-wire" fence. Mousetraps and "hot-wire" fences deliver remote punishment and are much more effective than interactive punishment since the pain is directly related to the behavioral act and the escape behavior is punished instantaneously.

In the case of George, extinction, punishment, and counterconditioning were all used. It was important that when the owner realized the critical point that the dog must not escape again, progress was made, even with a diehard escapee such as George (Case 7–2).

Tranquilizers. As we have already seen with fearful dogs, a behavioral problem can be so extreme that it becomes impossible to make contact with the dog for behavioral treatment. And there are escape artists like James (Case 7–3), that are so focused on escaping that they do not attend to anything else. Tranquilizing the dog can provide access to the dog so that he begins to notice the rewards that are available for good behavior.

CASE 7–2

History. Persistent escapes from enclosures was a specialty of George, a mixed German Shepherd. Over a period of time he broke loose from increasing levels of restraint, initially breaking through a leather collar, then a choke chain, and later escalating to scaling a 6-foot fence, sneaking under a cyclone fence, and chewing through all fence repairs.

Diagnosis. Escape Behavior

General Evaluation. George has been rewarded by successfully escaping. It is important that he no longer succeed in escaping and that he be rewarded for good behavior.

SPECIFIC INSTRUCTIONS

1. Adopt a routine where George is to stay in a kennel for 8 hours each day, the same time each day, even on weekends. His enclosure must be absolutely escape-proof.
2. Use a livestock "hot wire" electric fence if necessary. Locate the wire at a level on the fence that George must touch when climbing the fence (the back feet must be on the ground during contact).
3. Starting one week later give George a knuckle bone to chew on when you put him out each day.

CASE 7–3

History. During the past year James, a German Shepherd, has become a feverish escape artist whenever the owners are gone. When he began jumping the 7-foot fence the owners began leaving him in the house during their absences. He also barks frequently when put in the kennel.

Diagnosis. Escaping, Barking

General Evaluation. James will be tranquilized to reduce the level of misbehavior. At the same time a program of counterconditioning and remote punishment will be utilized.

SPECIFIC INSTRUCTIONS

1. Tranquilize James with Acepromazine, 10 mg tablet, 3 times a day, for 1 week. Double the dose once if needed.
2. Withhold affection from James when he is inside.
3. Put James outside 4 times a day. Turn on an outdoor radio. When James has been quiet for 5 to 15 minutes, go out and praise him. Over several days, increase the required quiet time for a reward.
4. For persistent escape attempts use remote punishment, such as squirting him with a hose (you are unseen) during escape attempts.

Chapter 8

Elimination and Urine Marking Problems in Dogs

This is an area of animal behavior where much misunderstanding arises. When we housebreak a dog, we are not actually teaching it eliminative habits, but simply relying on the dog's natural eliminative pattern of keeping its nest or den clean. In fact, many domestic animals, including swine, horses, and cats, will eliminate in selected areas.

NORMAL ELIMINATION BEHAVIOR

The reason most dogs are naturally fastidious about keeping their home area clean stems from the innate behavior of their ancestor the wolf. The offspring of parents that were the most fastidious in keeping the den clean of their own feces, which undoubtedly carried parasite eggs, survived in higher numbers than the offspring of parents that did not have such sanitary habits. Hence, the fastidious eliminative behavior has been "wired in" through the evolutionary process. The increasing use of drugs for treating parasites and disease agents allows relaxation of fastidiousness in dogs as the evolutionary process has continued in recent years.

Cattle, sheep, and chickens do not have this innate tendency and it would be very difficult to housebreak them. The same is generally true of primates. There are reports of people training chimps to use a "litter" box or a toilet, but they find the training laborious.

The ease with which dogs are toilet trained makes them favored pets. In fact, a great many dogs are probably toilet trained regardless of their owner's attempts to assist in the process.

Toilet training is a classic area for people to be anthropomorphic. When "accidents" occur we assume the dog knows it has been bad and we proceed with punishment accordingly. With our children (or other adults), we almost always take the time to verbally explain the particular acts that we are punishing. The person is told exactly what he has done wrong. But there is no comparable way to explain to a dog that has defecated in the bedroom why it is being punished. Some of the common punishments of dog owners, such as rubbing an animal's nose in the soiled area, or pointing to a mess several hours old, hinder the process rather than help.

THERAPEUTIC APPROACHES TO ELIMINATION PROBLEMS

The problems and concerns with eliminative behavior involve (1) the initial house training of a puppy or older dog, (2) house soiling in an older dog that seems to have lost its initial training, (3) submissive urination, and (4) urine marking in the house by adult dogs. Each of these problem areas requires a diagnosis of the specific problem and a specific therapeutic approach.

Initial House Training

It is important to remember that house training is a simple extension of a dog's innate behavioral tendencies, which enable it to generalize that the entire house is its den. There are genetic differences in the ease with which dogs of different breeds can be house trained. This is taken into account in Chapter 21.

The tendency of dogs to keep their nest clean is evident as soon as they are capable of leaving the nest (Ross, 1959). In the newly adopted puppy, this behavior is seen when it is placed in a small pen. If it eliminates at all, it will generally be on the side of the pen away from the bed.

From the puppy's standpoint, it is very difficult to perceive the entire house as the den. If permitted to explore the house, puppies naturally consider the living room and dining room as logical toilet areas. Toilet training should begin by restricting the animal to one small room or a part of a room, thus making the task simpler. The puppy should be taken outdoors frequently, whenever feasible, especially when the tendency to eliminate is high, such as after consuming a meal or awaking from a nap. If these training procedures are followed faithfully, the puppy will not eliminate at all in the small training area and will tend to hold urine and feces until it has the opportunity to eliminate outside. The owners should adhere to a regular schedule of taking the puppy outdoors until the puppy learns to keep the small area clean. Then they may allow the puppy some freedom in the house, while they watch it closely. In addition to allowing the puppy to gradually develop the concept of the house as the den, there is a visceral training of the intestinal system, much as there is in people, so that elimination will tend to occur at certain times of the day.

If a puppy cannot be taken outdoors frequently, it is advisable to first train it to eliminate on newspapers placed inside the small enclosure opposite the sleeping and feeding area. The enclosure might be made with a length of 2-foot chicken wire. When the puppy is accustomed to the routine of eliminating on the newspapers, it can then be allowed additional space within the house, providing that it continues to use the newspapers. By taking a series of steps at roughly 2-week intervals, the puppy may eventually be allowed access to all parts of the house. The newspapers should be kept in the same location, and when they are changed, the bottom papers should be placed on top of the new ones to provide olfactory as well as sight cues for the puppy.

Later, when it becomes possible to take the dog out on a regular basis, house training may be continued to encourage the dog to eliminate outdoors rather than on the newspapers. This requires taking the dog outdoors when the likelihood of elimination is high.

Although interactive punishment has traditionally been a part of housebreaking puppies, it does little to facilitate the process. When punishment has been useful, it was probably through creating a type of conditioned anxiety to house soiling if the animal was actually caught in the act.

If a dog has repeatedly soiled a particular area, it is sometimes useful to use aversion conditioning by tying the animal next to the place for several hours or even all day. This can be effective in causing the area to become aversive enough that the dog avoids it.

House Soiling in Older Dogs

Sometimes dogs who have been well house trained begin eliminating in the house, and there is no logical explanation for the change in behavior. At other times, the reason is quite clear, as with uncontrollable diarrhea. This disturbs the normal visceral rhythm for defecation and may force the dog to soil the house, confusing the dog's concept that the entire house is a home area to be kept clean.

A common problem is the dog that cannot make it through the night without messing on the living room or hallway carpet. The first task is to find out if the dog has the ability to make it through the night without urinating or defecating. This can be quickly explored by tying the dog very close to its sleeping area (this might be the owner's bed) for a night or two. If the dog has an "accident" in the night under these circumstances, some training of the dog's intestinal system will be necessary. This is an unpopular therapeutic approach, as the owner will need to get up at midnight or so to take the dog outdoors. The midnight excursions may be gradually advanced, over a period of a couple weeks, to the early hours of the morning. Successful training or retraining of the intestinal system should be possible if there is no persisting gastrointestinal disorder.

After the dog is able to make it through the night while restrained close to the sleeping area, it can be taken out each morning when the owner awakes at his normal time, and then rewarded with praise and food. Some owners may want to encourage their dogs to eliminate in a particular part of the yard at this stage of the training. There is some tendency for dogs to return to the same toilet area, such as a gravel run, and an owner may as well have some say in determining what area this is.

Another form of retraining is illustrated in the case of a Collie that could, at one time, be left in the house with another dog throughout the day while the owners were away. However, it suffered a bout of severe diarrhea, when it obviously could not control its elimination, and then its fastidious behavior did not return even after the problem with diarrhea had been completely resolved. The owner would frequently have to contend with a pile of feces on the kitchen floor. The situation was not improved by warnings, threats, or punishment. The clue that the problem was resolvable was that the dog almost always eliminated in the same spot in the kitchen. It was still basically toilet trained! For this problem, advantage was taken of the fact that dogs will rarely eliminate in an area where they sleep (Hart, 1979d). After cleaning the soiled area thoroughly, the dog's throw rug that it usually slept on was placed over the area. To prevent the dog from developing a new toilet area on the other side of the kitchen, the owner was instructed to tie up the dog so it could move only a short distance from the throw rug while still sleeping comfortably with access to food and water.

It was assumed that the dog's intestinal system had probably been conditioned so that there was a strong urge to defecate in the middle of the day rather than evening or morning. To solve this problem, a gradual approach was taken in which the owner was instructed to make special arrangements to return home at noon for about a week, so that the dog could be taken outdoors to eliminate. If the dog did go outdoors, it was rewarded with praise and food treats. The following week, the home visit was advanced to 1:00 p.m. and a few days later to 2:00 p.m. This procedure relied on the dog's natural tendencies to hold off elimination as long as possible, particularly since he was tied closely to his sleeping area. As the dog was left for increasingly longer periods of time, training of the intestinal system to a new schedule was gradually accomplished. In time, the owner was able to come home late in the afternoon, and finally after a normal work day. The second phase was directed toward gradually allowing the dog full access to the kitchen and later the rest of the house. The dog was first given an extra foot of rope during the day, while the owner continued to arrive home promptly after his working day to take the dog outdoors. The owner was instructed to give the animal a longer and longer rope until she felt ready to allow the dog freedom through the kitchen and eventually the house. Client instructions for this dog, Preston (Case 8–1), are included here.

This retraining program was probably overly cautious. The intestinal system, for example, might have progressed more rapidly than in increments of one hour, and the very gradual increases in the length of rope were probably more cautious than necessary. However, as in many

CASE 8–1

History. The owner of a 5-year-old male Collie, Preston, reports that her dog was well housebroken until 6 months ago, but after a bout of serious diarrhea the dog cannot be left in the house during the day without defecating. The dog continually soils the same spot in the kitchen.

Diagnosis. Problem Elimination

General Evaluation. Preston is still toilet-trained. The goal is to retrain the dog to use a new location.

SPECIFIC INSTRUCTIONS

1. Change Preston's soiling area into a nest. Thoroughly clean the kitchen and cover the spot with Preston's sleeping rug. When Preston is indoors, confine him on a short lead to the sleeping rug.
2. For the first week, take Preston outside to eliminate at midday. If he does, reward him with praise and food treats. In each successive week, take him outside an hour later. Finally, Preston should go the entire workday without going outside.
3. Then, extend the length of Preston's rope by one foot every three days until he has free run of the house.

behavioral problems, reducing the chance of further accidents sometimes requires that we spend more time in resolving the problem than might otherwise be necessary.

Submissive Urination

When a dog is urinating inappropriately, it is important to notice whether the symptom is a fear reaction. Since the client may not know the dog is responding out of fear, he may make the problem worse by viewing it as misbehavior. With a few well-directed questions about the occurrence of inappropriate urination, you can determine if it is a submissive reaction. This problem is also addressed in Chapter 4 on fear reactions.

This problem stems from the use of urine as a submissive signal and thus cannot be approached in the same way as toilet training or retraining. Submissive urination occurs during emotional situations when a dog becomes anxious or nervous, and it releases urine as part of the submissive gesture. In the wild, the urination and other submissive signs have the effect of suppressing or inhibiting an aggressive attack by another dog. Since submissive urination is displayed mostly by puppies, owners sometimes confuse it with inadequate toilet training. In older dogs this type of submissive behavior occurs in females much more than males.

When we attempt to punish submissive urination, the puppy's anxiety increases, which causes it to become more submissive and urinate more, because the threat of physical attack seems even more imminent. The dog is attempting to turn off our aggressive behavior by further urination. Increasing punishment makes the behavior worse, and the dog never learns what it is doing wrong. This type of urination problem should be handled by creating a situation where the animal is more relaxed and less anxious or nervous, as we attempt to desensitize the behavior.

In general, two main principles are involved. The owner should refrain from any further interactive punishing of the dog for urination. Instead, attempts must be made to avoid circumstances where the behavior is most likely to occur; this is usually in greeting or calling the dog. In high-risk areas, owners should ignore the dog to reduce the chance of urination. All petting and fondling should be done at a low-risk time, such as after one has been in the animal's presence for 10 to 15 minutes. If the dog does urinate, the owner should simply walk away in disgust.

Some training sessions may be arranged to remotely punish submissive urination. In a high-risk situation, it is prearranged to have an accomplice wait with a water sprayer in a position to squirt the dog without being seen. One person approaches the dog with hands to the side. When, as expected, the dog urinates, the accomplice squirts the dog. The other person then calmly turns and walks away. In this situation the remote punishment is not connected with the person the dog is approaching, therefore, the punishment should not evoke further urination but rather disrupt the urination. Review again the sample case of Shabby (Case 2–1, pg. 23).

In further training sessions when the dog approaches, or is approached, but does not urinate, it should be mildly praised and rewarded with food. The food reward also produces an emotional state incompatible with the anxiety leading to submissive urination.

Urine Marking in the House

Owners of male dogs are well accustomed to seeing them urine-mark many trees and posts along the way whenever they take a walk. Sometimes a dog begins to also lift his leg inside the house and mark a favorite chair or table leg.

History. During puberty most male dogs begin to use the leg-lift posture to urinate and frequently mark vertical objects in their environment, apparently as an expression of territoriality. The behavior is often evoked by the presence of an intruder male dog or another dog walking down the street. The behavior is sexually dimorphic in that normal females rarely use the full leg-lift posture, and they urinate infrequently in large quantities rather than frequently in small amounts as do males. Early androgen treatment of females can induce them to display the male type of urination (Beach, 1974). Natural androgenization of females in utero by existing adjacent to male fetuses may explain why some presumably normal female dogs do display leg-lift urine marking behavior later as adults (see Chapter 23).

Male dogs that are house trained usually do not urine-mark in the house. This makes sense because in the wild it would obviously be disadvantageous for males to deposit urine in a den around puppies. Experimental evidence shows that the urine marking tendency is much higher in novel areas than in a dog's home area (Hart, 1974a). But on occasions the drive to mark in the house, evoked by an excitatory stimulus such as the presence of a new dog in the home or neighborhood, may be so great that even a well-housebroken dog may start urine marking household furniture, draperies, or walls.

Castration has been found to suppress the behavior in gonadally intact males about 50 to 60% of the time. For an intact male, castration is a first step in correcting urine marking, as in the case of Prince (Case 8–2). Interestingly, castration does not appear to markedly affect urine marking in novel or outdoor areas (Hart, 1974a). Apparently, outdoors, where the olfactory and visual stimuli evoking markings are very strong, castration has little effect.

When castration is not effective, or when a castrated male takes up urine marking in the house, the administration of a long-acting progestin is indicated. The injection of one dose of medroxyprogesterone at the rate of 5 to 10 mg/kg, as outlined in Chapter 23, page 245, may control the behavior in many of these problem dogs. One or two repeat injections at 1- or 2-month intervals may be necessary.

CASE 8–2

History. This intact 6-year-old, male Miniature Schnauzer, is urine marking in the house. Prince has two favorite marking sites that he has been frequenting during the past two months.

Diagnosis. Urine Marking in the House

General Evaluation. Castration may solve this urine marking problem. If not, there are other behavioral and pharmaceutical options.

SPECIFIC INSTRUCTIONS

1. Castration has a 50% chance of eliminating this behavior.
2. A change in urine marking behavior after castration may be rapid or gradual over 3 months.
3. If marking persists after 3 to 4 weeks, you may consider an injection of a progestin.
4. Behavior modification approaches can be used to protect the favorite marking sites. Set about 4 mousetraps at the sites so that Prince will be caught and punished in the act of marking. Protect the cat from being caught in the traps.

Chapter 9

Feeding and Related Problems in Dogs

Owners are often concerned about the eating habits of their dogs. For some finicky dogs, finding an appealing food may be a problem. Other dogs eat well for a while and then seem to stop eating for no apparent reason. Anorexia may be secondary to a systemic disease, or it may be the major problem. Obesity is also a common problem, and although dog owners may control their dog's access to food, they seem to have little success in reducing the animal's weight. Another eating problem, referred to as pica, occurs when dogs consume inappropriate objects such as pieces of wood or rocks. Finally, coprophagy, consumption of cat or dog feces, is especially objectionable to some dog owners.

NORMAL FEEDING BEHAVIOR

Many people believe that a natural diet for carnivores consists solely of meat. It is a dangerous misconception, however, to believe that dogs can live on meat alone. The diet of wild canids and felids consists of the entire carcass of rodents and larger mammals, as well as occasional vegetative matter. Therefore, skin, bones, and the contents of the intestinal tracts of herbivorous birds and mammals constitute a good part of the diet, whereas a package of meat has less nutritional variety.

In a small way, dogs may attempt to supplement their ongoing food supply by a behavioral pattern that could be called hoarding or food storage. A common example of a behavior unique among the domestic species is the tendency of some dogs to bury bones. Even when some dogs are in plastic or metal cages, they attempt to cover a bone by using papers on the cage floor. Wild canids, including wolves and some species of fox, bury small prey and may go back to the source of the cache during lean months of the year. It is common for dogs to bury bones but not to dig them up later, so the practice may not be meaningful in supplementing the food supply. However, some dog owners claim to have seen their animals bury small animals such as woodchucks and later uncover them after the tough skin had decomposed.

Social and Environmental Influences

Like their wild ancestors, dogs tend to eat rapidly. Of course, when carnivores kill prey, there is competition for food and the animals that eat rapidly receive the most food. The wolf has a remarkable ability to gorge itself, and it has been reported that it can consume up to a fifth of its own body weight in meat. Some dogs can also consume enormous meals; a male Labrador Retriever was reported to consume 10% of its body weight in canned food (Mugford, 1977).

Social facilitation can increase eating in dogs, particularly puppies, when they are fed a few times a day. Puppies in one study ate 40 to 50% more when they were fed in groups than when they were fed alone (Ross and Ross, 1949), but these effects are temporary. Over time, group-fed puppies probably eat no more than those fed alone. The effects of social facilitation are also less apparent when food is readily available all day. When adult Miniature Poodles had continuous access to a complete semimoist dog food, there was little evidence of social facilitation (Mugford, 1977). The dogs tended to eat frequent, small meals, mostly at night.

If food is restricted, however, dominance interactions, especially among puppies, occur during feeding. The dominant animals may get such a large proportion of the food that the more subordinate ones are undernourished. Feeding puppies with several pans of food is a way of preventing this problem.

Familiarity with the environment naturally plays a role in determining a dog's appetite. An animal hospital is not a particularly good place in which to evaluate a dog's eating behavior, because the strange environment may suppress eating. The hospital diet may be different from the dog's customary diet, and the time at which the dog is fed may also be different.

Most dogs adapt to a new feeding regimen surprisingly well. One study of eating behavior in hospitalized dogs found that within 3 days most dogs adapted to the hospital regimen, as indicated by how rapidly they consumed their meal after being fed (Boulcott, 1967). Further evidence showed that dogs ate more in the hospital than at home. Although healthy dogs may adapt to a hospital environment, there is evidence that several factors affect a dog's appetite, and sick or convalescing dogs could suffer if the hospital staff rigidly adhere to the principle that a dog will eat if hungry enough.

Food Aversions

A canine's likes and dislikes can reflect acquired aversions as well as acquired tastes. Animals may develop an aversion to a food that has made them sick or nauseous. Rejecting foods that have been paired with sublethal doses of x-rays, lithium chloride, or apomorphine is an obvious example of acquired aversion. These various treatments cause post-ingestional nausea and sickness, and the taste or smell of the food takes on aversive properties after one or more pairings with the illness-producing treatment (Garcia, Hankins, and Rusiniak, 1974).

A striking example of aversive conditioning in canids is the attempt to produce an aversion in coyotes to sheep (Gustavson, *et al.*, 1974) as a possible approach to control sheep predation. The investigators wrapped lamb meat, laced with lithium chloride, in a lamb's wool covering and allowed coyotes access to it. After one or two pairings, the coyotes, who were known sheep-attackers, no longer chased sheep in the experimental enclosure. The investigators even related seeing one of their coyotes being chased around the pen by a lamb. The specificity of the conditioning is illustrated by the fact that the coyotes would readily go after rabbits, even though they reportedly had an aversion to sheep. Aversions are most easily produced to novel foods; familiar tastes seem to be resistant to association with illness. The ability to associate a food with gastrointestinal illness seems to diminish once the animal makes the association that its flavor is safe or its flavor makes it feel better.

The function of conditioned food aversions in nature is probably to protect animals from repeated ingestions of food that produce gastrointestinal illness. Animals have food allergies and the development of aversions would lead them to avoid attempting to eat the food that produces illness.

One would think that the principle of conditioned aversions might provide an approach for treating some types of undesirable ingestive behaviors, such as coprophagy in dogs, or even predation on birds and cats. However, animals are often familiar with the material to which we would like to produce an aversion, and dogs seem to resist associating familiar tasting foods with illness.

Attempts to produce food aversions in dogs have met with little success. This may result from the fact that aversions are more difficult to induce with familiar rather than novel foods; also, the years of domestication may have made dogs less easily conditioned to food aversion. In wild species, in which aversion conditioning has a definite adaptive role in protecting animals from toxic foodstuffs, the behavior is maintained by natural selection.

An important concern is producing an aversion unintentionally by administering drugs to animals in a small amount of food (the effects of this are not well studied). There are some experimental results using common drugs that bear on this practice. If drugs such as sedatives or tranquilizers are given to rats just after they have consumed milk, a conditioned aversion is formed to the milk (Berger, 1972). These observations support the possibility that giving drugs in food could produce an aversion to the food, especially if the procedure is repeated several times with the same food.

PROBLEMS WITH FEEDING BEHAVIOR

Pica: Ingestion of Inappropriate Materials

Excluding toxic substances, dogs are not necessarily careful about what they chew and swallow, and sometimes they consume dirt, gravel, and pieces of paper along with the food. Occasionally, dogs selectively consume inappropriate materials such as rocks or pieces of rubber intentionally, even to the extent of causing stomach or intestinal impaction.

The cause of this behavior is often unclear. One cause of the behavior may be that it is attention-getting and the problem can be treated as other attention-getting behavior, with extinction and remote punishment. Types of punishment that have been used with some success include baiting the objects with a hot pepper powder, mousetraps, or hot wires. Treating the problem like destructive behavior and using departure training sessions is another approach if pica occurs daily.

Both dogs and cats have been observed eating grass, especially long blades. This phenomenon has not been studied experimentally, so nothing conclusive can be said about the reasons for eating grass or its physiological effects. Some dog owners assert that their dogs eat grass when there is an intestinal disorder, and that they often vomit after eating the grass. Other dog owners report that eating grass is followed by a loose bowel movement, suggesting that the grass has a laxative effect. It could be argued that the adaptive value of eating grass may be to act as a physical purge for intestinal parasites.

Coprophagy

There is no recognized uniform explanation of this behavior, but some possibilities have been suggested. Both male and female dogs are innately programmed to consume fecal droppings of puppies in the nest or whelping area. Perhaps when a dog is bored, coprophagy strikes it as a natural pastime. In some instances, eating feces can be a type of attention-getting behavior. Owners may react emotionally to the sight of their dogs going after feces, and a dog may pick up this reaction as a means to garner additional attention. Dogs that habitually consume their own feces are often confined dogs that receive little human attention. One might also guess that dogs could be attempting to satisfy some unknown nutritional deficiency.

The behavior of a dog consuming its own feces or those of other dogs or cats is certainly maladaptive, since the behavior would tend to infect or reinfect dogs with parasite eggs. Coprophagy in very young animals may serve the purpose of establishing an appropriate intestinal flora, but this potential benefit must be weighed against the disadvantage of consuming parasite eggs.

Applying a repellent to the feces is the most frequent approach for eliminating objectionable coprophagy in dogs. A foul tasting or smelling substance, such as hot pepper powder, sometimes works, as in the case of Harry (Case 9–1).

Anorexia

It is wise to look for a medical or physiological cause of loss of appetite. Does the dog have a gastrointestinal disorder? Is he suffering from a diagnosed or undiagnosed food allergy? If an allergy is known, simply eliminating the allergen may not cure the problem if the food still has the same general olfactory and taste characteristics. The anorexia could be an aversion conditioned by the gastrointestinal sickness produced by the allergen. It may be necessary to change the characteristics of the food to side-step the aversion.

Anorexia is a normal response to many sicknesses in general. By losing its appetite, an animal in the wild will stay holed up in its den conserving heat to help maintain an elevated body temperature (fever) to combat bacteria or viruses. The best therapeutic approach here is to allow the anorexia during the height of the illness and then help the animal regain normal appetite in the convalescent stage.

When the loss of appetite is due to inability to smell, secondary to a severe upper respiratory disease, placing food into a dog's mouth causes taste receptors to be stimulated and may interest the animal in eating again. Sometimes dogs may stop eating for emotional reasons. We can also take advantage of the appetite-stimulating side effects of steroids, including corticosteroids and long-acting progestins. Progestins such as

CASE 9–1

History. Harry, a castrated male Cocker Spaniel, is being treated for a variety of behavioral problems, including aggressiveness, urination, and destructiveness. Although those behaviors have diminished with treatment, he has now added the problem of coprophagy.

Diagnosis. Coprophagy

General Evaluation. Harry's new routine seems to be related to the new problem of eating his own feces. Applying a taste repellant to the feces may curb this habit.

SPECIFIC INSTRUCTIONS

1. Continue feeding Harry twice a day.
2. Apply cayenne pepper to his feces. His mouth will burn if he ingests the feces. Stage a situation where some feces are left in an obvious place for Harry. Apply cayenne pepper under the surface throughout the feces. Attempt to observe Harry's reaction when he eats the feces and note if he is repelled.
3. If Harry was repelled by the hot feces, keep all feces away from him for a week and then stage another test, again using the pepper powder.
4. Continue with test sessions until you are certain Harry is avoiding feces. If he later starts eating feces again repeat the above process.

medroxyprogesterone (Depo-Provera, 5 mg/lb.), which are used for objectionable male problem behaviors, have pronounced appetite stimulating effects. This approach was utilized once on a male dog that became disturbed when the young boy with whom the dog associated daily after school lost interest in the dog and began playing with other boys. The dog became depressed and stopped eating. The appetite was markedly improved the day after the administration of a long-acting progestin. Frequent attention thereafter from a young girl in the family rehabilitated his emotional state. If dogs are avoiding food and acting anorexic because they have learned it results in receiving more attention, we should treat the problem as attention-getting behavior, as described in Chapter 6.

Obesity

Most animals maintain a normal body weight with no apparent effort. It seems they have a remarkable ability to adjust their consumption of calories in terms of carbohydrates, fats, and protein to meet their energy requirements. Furthermore, animals force-fed extra food or given calorically diluted food can adjust their food consumption accordingly. Yet, obesity sometimes arises as a problem. Certain disease conditions may lead to obesity. These are hypopituitarism, tumors of insulin-producing cells, excessive production of adrenal corticosteroids, and cranial tumors producing pressure on the ventromedial hypothalamus.

Obese dogs take in too many calories for their energy expenditure. The main reason pet dogs overeat is probably because they are offered an excessive amount of highly palatable food, and they also lead a rather inactive existence. Like wolves, dogs are capable of consuming large amounts of meat or animal products at one time, considerably more than they need on a daily basis. This inherited characteristic related to the irregular availability of prey for wild canids can lead to repeated overeating when appetizing food is continuously provided.

Contrary to what we might believe, dogs seem to prefer a diet containing sucrose to the more bland diets (Houpt, *et al.*, 1979). Dog owners who treat their dogs to cookies and candies may be contributing to their pets' obesity.

Aside from questions of owner-inspired obesity, the possibility of a genetic predisposition toward obesity should be given consideration. Some animals may have inherited a tendency to be fat and may have an appreciably larger number of fat cells than other animals. The number of fat cells an animal has tends to persist throughout life, and the number of fat cells tends to set the level for total volume of body fat. Some theorists claim that an animal's nervous system tends to "defend" the size of these fat cells. Thus, to force an overweight animal to reduce could be comparable to starving an animal with a more normal number of fat cells. This could bring on such signs as irritability, depression, and inactivity.

There is also the possibility that pet owners, in their concern to keep young puppies or kittens as healthy as possible, have supplemented the mother's milk to an excessive degree, causing hyperplasia of fat cells in the infants and promoting a degree of permanent obesity in an adult dog. In instances of genetically induced obesity or over-feeding as an infant, one could take the position that we might as well forget trying to get the owners to make the dog reduce.

It is generally believed that castrated or spayed dogs have a higher tendency to become fat (and lazy) than animals with intact gonads. From experimental work on the female rat, there is no question that ovariectomy leads to increased body weight, food intake, and fat levels (Leshner and Collier, 1973; Wade, 1976). Administering an estrogen reverses this effect and results in a reduction of food intake and body weight. A study by Houpt *et al.* (1979) has shown that ovariectomy in female dogs caused a gain in body weight and food intake for at least 15 weeks after surgery over that of sham-operated control dogs. However, the difference between spayed and sham-operated dogs was only 2 to 3 pounds, and the increase in body weight was not reflected in increased deposition of subcutaneous fat. These observations suggest that a slight increase in body weight could be attributed to hormonal changes following ovariohysterectomy, but any major changes are due to dietary or exercise management.

The experimental findings pertaining to the effect of castration and food intake in males are equivocal. For example, male rats that are castrated undergo a moderate reduction in both food intake and body weight (Gentry and Wade, 1976). Much of the weight loss is due to reduction of muscle which is replaced by fat tissue (Mitchel and Keesey, 1974). As with female dogs, we must, for now, conclude that the effects of a gonadectomy on weight gain are probably minor, and that the occurrence of obesity in a castrated male is most appropriately attributed to the owner's behavior in simply feeding the animal more than it needs for its energy output.

Chapter 10

Sexual Behavior Problems in Dogs

There are two types of clinical problems relating to sexual behavior: One is sexual inadequacy and lack of interest on the part of both males or females to mate. The second type is when dogs, especially males, mount people, inanimate objects, and other dogs to an excessive degree. An understanding of normal behavior is needed to deal with both types of problems, so this section includes a brief description of normal sexual behavior of males and females before going into problem areas.

NORMAL SEXUAL BEHAVIOR

Male puppies often exhibit sexual mounting, even with pelvic thrusting. Such mounting is a normal part of play behavior and is, in fact, necessary for the development of normal sexual responses in adults. If puppies are raised in isolation from other dogs and do not experience sexual play, as well as other social interaction, they may, as adults, experience difficulty in executing copulatory behavior. The most noticeable defect is an improper orientation of mounting toward the genital region of females (Beach, 1968). As puppies approach adulthood they have less interest in mounting other males and channel their sexual responses toward females in estrus.

Adult mating behavior generally begins with anogenital investigation of the female (Fig. 10–1*A*), some attempts at playing with her, and a few mounts with pelvic thrusting. When genital intromission occurs, it is somewhat a result of trial and error thrusting. Thus, a male may mount and thrust several times before achieving intromission (Fig. 10–1*B* and *C*). The female aids in intromission with her receptive responses, such as the lateral curvature of the rear quarters, deviation of the tail, and movement of the external genitalia.

Male dogs differ in the extent of their courtship patterns, with some attempting copulation almost as soon as they are placed with females, and others taking longer. These individual differences seem to reflect

Fig. 10–1. Mating behavior of dogs. The female displays lateral curvature and tail deviation as the male investigates the genital area *(A)* and mounts to begin pelvic thrusting *(B)*. Upon intromission *(C)*, the male engages in intense pelvic movement and leg stepping for 15 to 30 seconds (referred to as the intense ejaculatory reaction),

Fig. 10–1. ***Continued*** which is often terminated by the female throwing the male. *(D)*. Experienced males usually dismount before they are thrown. The female's excitement often continues briefly after the male is dismounted *(E)*. The genital lock continues for usually 10 to 30 minutes, and during this time both dogs are quiescent *(F)*.

contrasts in genetic background and do not reflect the degree of sexual drive or motivation (Hart, 1968).

At the time of genital intromission, a marked change occurs in the behavior of the male. As the front legs are pulled caudally, the tail is deflected downward, and stepping activity of the rear legs is initiated along with intense oscillations of the pelvis. Some animals show alternate stepping of the back legs so high that they are thrown off balance. This behavior, termed the intense ejaculatory reaction, lasts for usually 15 to 30 seconds. At this time the sperm-dense fraction of semen is expelled. At the time of ejaculation, the penis engorges within the vagina and the male and female are effectively locked together. The bulb of the penis is so engorged that it must remain in the vagina until erection subsides.

During the initial part of the intense ejaculatory reaction the female usually stands rigidly. Toward the end of this reaction the female often starts twisting and turning, and she may throw the male before he dismounts (Fig. 10–1*D* and *E*).

The male usually turns and lifts one leg over the penis after dismounting so that, following a successful intromission, the two animals end up in the tail-to-tail genital lock position (Fig. 10–1*F*). Typically, the lock lasts from 10 to 30 minutes, although from 5 to 60 minutes is within a normal range. Prostatic fluid continues to be ejaculated during the lock.

Obviously, the lock is not necessary for fertilization to occur, since litters can be produced by artificial insemination. While it was once believed that contraction of the female's vaginal constrictor muscles maintained the male's erection during the genital lock by occluding the venous return from the penis, it is now known that complete erection of the penis is a manifestation of a spinal reflex in the male. During the reflex, some penile muscles contract and occlude the venous return while others pump arterial blood into the erectile tissues (Hart, 1970; 1972). If the penis becomes engorged prior to intromission, or if it is pulled from the vagina prematurely, detumescence can be facilitated by touching the coronal edge of the glans just behind the urethral process.

Sexual Behavior of Females

Prior to puberty females display scarcely any sexual responses. The ovaries of females reaching puberty secrete estrogen twice a year. Estrogen has the effect in most animals of increasing activity. Thus, a female dog usually moves about more, vocalizes more frequently, and may act nervous during estrus.

Dog owners report that the urine and vaginal secretions of female dogs in proestrus and estrus are attractive to males. Male dogs are reputedly attracted from some distance away to the vicinity of an estrous female, presumably by one or more chemical attractants in the urine. Because they communicate the message of sexual receptivity, these sexual attractants are sometimes termed sex pheromones. These pheromones are probably not so potent as to be detected by males from miles away, but they are certainly noticed by males on their neighborhood rounds.

The leg-lift urination posture is displayed by some female dogs when they are in estrus. This probably functions to assure that the urine is deposited on prominent (vertical) objects in the environment. The nature of the pheromone of female dogs is not known.

Female dogs that are in estrus may evoke mounting from both male and female dogs. Conversely, when placed with a male dog, an estrous female may mount the male, especially if the male's sexual advances are too slow. Dog owners with two female dogs may occasionally even observe a spayed or anestrous female mounting an estrous female or vice versa. Since this behavior is relatively normal, it is not properly referred to as homosexual behavior.

PROBLEMS WITH SEXUAL BEHAVIOR

Lack of Sexual Interest in Breeding Animals

This problem may be of a temporary or permanent nature. One type of temporary problem relates to the need some males have for familiarity with the environment before mounting. For this reason, females are usually brought to the male's environment for breeding. However, if a male dog readily adapts to new environments, there is no reason that he could not be brought to the female's home.

Lack of correct orientation of mounting may be a result of too little sexual play and contact with other puppies during early life. This is a relatively permanent problem and may not be amenable to adaptation or behavior modification.

Females may show a preference for certain males even when they are in full estrus. A less frequent occurrence is that some male dogs prefer to mate with certain females and tend to reject other females that are fully receptive. In modern breeding operations we expect male dogs to be promiscuous breeders as are farm animals. However, the wild ancestor of the dog, the wolf, is basically a monogamous species. Thus, dogs are genetically predisposed to be more selective than breeders would like.

The term silent estrus refers to the condition in which a female shows the physiological but not the behavioral signs of estrus. Presumably the brain is not responding to estrogen secretion the same as the genital system. One precaution to keep in mind before diagnosing this condition is that female dogs have been observed to show definite preferences for males, and they may appear to be in silent estrus when in fact in the presence of a more desirable male they would show sexual interest.

Artificial Collection of Semen

When problems with breeding dogs become too difficult to handle behaviorally, artificial insemination is frequently employed. If it is necessary to obtain semen from male dogs for artificial insemination, remember that the ejaculatory reflex is very susceptible to supraspinal inhibition, and this inhibition must be somewhat overcome before the reflex can be manually activated. In some instances, one may need to place a receptive female, perhaps a spayed female injected with estrogen, close to the male. To elicit ejaculation, the body of the penis proximal to the glans should be rubbed until rhythmic contractions begin in the striated penile muscle (bulbospongiosus muscle) surrounding the base of the penis as it leaves the pelvic cavity. Since there are no receptors for the ejaculatory reflex over the glans penis, manually stimulating the glans will not evoke ejaculation. Actually, an inhibitory effect on ejaculation and erection is caused by touching the coronal edge of the glans near the tip of the penis.

If rubbing the body of the penis proximal to the glans does not produce ejaculation, then additional stimulation may be useful. Continuing to stimulate the body of the penis while placing a finger on the tip of the urethral process at the same time will evoke ejaculation in most dogs that are otherwise inhibiting the ejaculatory response (Hart, 1970).

Objectionable Mounting Behavior

Male puppies may mount people, especially children because they seem to tolerate it. This is probably a reflection of puppy-play sexual behavior and most puppies can be expected to grow out of it. Punishment is usually sufficient to stop the behavior. Dogs that direct mounting activity toward people may have formed an overattachment to people during early life, especially if there was little opportunity for the dog to interact with other dogs. Punishment usually works here too.

Mounting behavior may also be displayed frequently if it is followed by attention. Thus, if punishment is used to stop mounting, the punishment should also be followed by social isolation, or at least indifference to the dog for an hour or two.

The mounting of people or dogs by adult dogs is most common in males, but may occur in females possibly as a reflection of fetal androgenization (see Chapter 23). Such mounting in males, or the onset of mounting in adulthood, is an indication for castration. Results of a clinical survey suggest that one can expect favorable results from this operation about 60% of the time, with rapid results about one-half of the time, and a more gradual cessation of mounting the other half of the time. Castrated males or spayed females that persist in mounting can be treated with medroxyprogesterone injections or orally with megestral acetate (see Chapter 23).

Chapter 11

Maternal Behavior Problems in Dogs

Professional advice is often sought by dog owners who want to understand what is normal for bitches when they become mothers, and also for treatment of abnormal or undesirable aspects of maternal behavior.

Maternal behavior is an area in which the behavior of the dam is so important to the survival of the offspring that evolutionary forces have programmed the neural circuitry for this behavior into the brains of dogs. Even a mother that has had no previous opportunity to observe or engage in the care of newborn suddenly, at the time of parturition, performs a relatively complex sequence of maternal tasks which continues and changes until her offspring are able to survive on their own. The specific elements of maternal behavior are performed with perfect timing and precision. The mother has an emotional attachment to her young that represents the strongest interindividual bond in nature. She is even willing to risk her life for these new animals which she hardly knows.

While maternal care is essential for the welfare of newborn, social interactions with the mother and littermates also prepare the young for later social behavioral patterns and temperament. These interactions are crucial for shaping the behavior of pups to be desirable as pets, and deprivation of this interaction, by early weaning or orphanage, is costly.

Many professional dog breeders and family breeders are all too anxious to step in and help canine mothers when any aspect of maternal behavior is insufficient, or the lives or comfort of the pups are threatened. For example, supplemental bottle feeding and assistance in weaning pups onto solid food are often provided, as well as help in the birth processes by cleaning and grooming the pups, cleaning the mouth, stimulating respiration, and inducing the first defecation. Although the innate aspects of maternal behavior are programmed into the brain, this programming is not immutable, and our intervention through centuries of dog domestication has facilitated survival and reproduction of mothers with defects in neural programming.

The gene pool of most breeds includes widespread defects in maternal responsiveness, which cause a great deal of variability in maternal behavior among individual dogs. Some dogs exhibit complete and effective maternal behavior from birth through weaning, and others are totally inattentive to their young. With some mothers, few or no pups would survive if assistance were not provided to bitches in giving birth and raising young.

Since relatively few of the behavioral patterns of mothering are learned, it is erroneous to assume that major deficiencies in maternal care stem from a bitch's inexperience, or to think that a bitch will learn new behavior with subsequent litters. Some basic habituations or adaptations, however, may play a role in making experienced bitches less nervous or anxious than naive ones. To the degree that nervousness interferes with the behavior of growing pups or leads to cannibalism, the experienced bitch may be a better mother.

Dog owners occasionally ask if having a litter of puppies will calm the mother. Although this effect may occur occasionally, perhaps by chance, there is no documentation to support this notion. It is certainly not advisable for a client to have a bitch bred to try to solve a behavior problem involving excitability or excessive activity.

There are two unique aspects of canine maternal behavior as compared with other mammals: One is that their wild ancestors were monogamous, and sires stayed around and customarily played a role in the care of the litter. Secondly, females that have not conceived may undergo mammary gland development, produce milk, and display maternal behavior at the same time of year as the bitch with offspring. The extended family of the wolf includes aunts who serve as nursemaids, the sire, and other male wolves of the pack who, along with the mother, often help at the time of weaning with feeding of the young by regurgitating food for their consumption. Of course, all members of the pack play a role in teaching wolf pups to hunt. Since pack members are usually related, they are contributing toward the survival of offspring with which they share some genes.

NORMAL MATERNAL BEHAVIOR

The following is a summary of normal maternal responses. It is obvious that one cannot really discuss "normal" canine behavior on the domestic scene. Even the female's role as the sole parent has stemmed from centuries of human modification and is highly variable.

Behavior Before and During Parturition

A pregnant bitch often becomes somewhat restless and nervous a few days prior to parturition. The dog may follow the owners around excessively, and may also tend to lie down for only a short while, get up, move, and then lie down again. Licking of the abdominal and genital areas increases just before parturition.

At the time of parturition, or labor, pups are delivered in four stages: (1) contraction, (2) delivery of fetuses, (3) delivery of the placentas, and (4) the intervals between deliveries. Uterine contractions begin in the first stage, and there is a good deal of straining. Most bitches lie down during contractions, and they may frequently move and change positions. Uterine and abdominal contractions become more intense in the second stage as the fetus moves rapidly through the birth canal. The female often breaks the fetal membranes with her teeth as the head or buttocks of the fetus appear at the vulva. This tugging on the membranes may actually pull the fetus through the birth canal.

After the newborn passes through the birth canal, the bitch consumes the fetal membranes and licks the newborn vigorously, which usually causes the first respiratory movements. The mother continues to lick and groom the newborn, and when the placenta is delivered, it is usually eaten also. The mother generally bites off the umbilical cord while she is eating the placenta. The mother's eating of the placenta and umbilical cord, with the associated pulling and stretching, seems to cause constriction of the vessels in the cord that is still attached to the pups. Occasionally, the cord is broken by movement of the newborn and mother. A neonate pup may sometimes be accidentally wounded when a bitch eats right into the abdominal wall after chewing off the cord. If the umbilical cord is not broken soon after birth, it may be necessary for a person to intervene at this point. During licking and grooming the mother concentrates more on the anogenital region of the newborn, stimulating defecation and the expulsion of meconium.

Between deliveries the mother continues to lick and groom the newborn animals as well as her own genital region, and also cleans the bedding that has been soiled with amniotic fluids. She may lick her own body more than that of the newborn. The licking seems to be stimulated by the presence of fluids rather than by the presence of the newborn.

The stages of labor vary widely in duration, since puppies may be born just a few minutes or several hours apart, but usually the total duration of parturition does not exceed 12 hours. Following the birth of the last pup, the female lies almost continuously with her young for a period of about 12 hours.

Nursing

Nursing is the focus of much of the interaction between a bitch and her puppies. This interaction begins as soon as the young begin to suckle, which can occur before the delivery of the last newborn. While finding a teat might appear to be a random process, it is more systematic than that. Moving toward the warmth of the mother's body, neonates crawl slowly and irregularly with paddlelike movements of the front legs while pushing with the hind legs. They "scan" the area ahead by moving their heads from side to side, and eventually they come into contact with the mother's ventral wall. The young then climb onto the mother's body and nuzzle into the mother's fur until they locate a nipple. In rats, olfactory cues emanating from the teat guide the young (Singh and Hofer, 1978), and it is certainly possible that the same is true for dogs.

The first stage of nursing lasts for 3 weeks after parturition, and through this phase the mother initiates essentially all nursing sessions. The bitch approaches the pups and lies near them, and then arouses them by licking. With time the newborn become very adept at finding teats and responding to the mother's solicitous behavior. Apparently, puppies do not take specific nipple positions.

In the second phase of nursing, beginning in about the third through the fourth to fifth week, the young are able to leave the bed. With their eyes and ears functioning well, they are able to recognize and interact with the mother outside the bed. Nursing sessions may occur inside the bed or outside and are now mostly initiated by the young. Usually the mother cooperates by immediately lying down, or by exposing her teats if she is already lying down.

The period from about the thirtieth day after birth and extending into weaning constitutes the third phase, when practically all nursing sessions are initiated by the young. Increasingly, the young follow the mother, and as time progresses, the mother begins to avoid nursing attempts of the young. She may lie with her mammary region against the floor or go away where the young cannot reach her. Near the end of this phase, when the mother is less accessible to the young, the young become capable of taking adult food.

In wild carnivores, where young must eventually learn to kill prey, the transition from nursing to solid food is gradual. Various strategies to facilitate the transition in the wild are used in different species. In wolves, the pack members, including the dam and sire, feed their pups by regurgitating or disgorging a freshly consumed meal. Wolf pups eagerly "beg" for the partially processed food by pawing and biting at the lips of an adult who has just returned to the nest. In time the pups are also introduced to freshly killed small game, or parts of a carcass of a larger prey, which are brought to the den. It is only when pups have acquired hunting skills by traveling with the older members of the pack that the pups are capable of living on their own.

Human owners have provisioned dogs at the time of weaning for centuries. Of particular importance is that the food for weaning puppies be palatable and easily chewed. Since their domestication, bitches rarely disgorge food to their young in the manner of wolves. The behavior is so sufficiently objectionable to humans that it has probably been selected against.

Adoption of Puppies

Unlike ungulates that reject alien young, a lactating bitch will usually accept them. The adopted puppies may even be a couple of weeks older than her own, or of another species, such as young cats, skunks, or squirrels. Remembering again a dog's ancestry, there is no particular advantage for a wolf bitch to reject another's young since they are probably also related to her.

Interactions Between a Mother and Her Young

Newborn puppies are in contact with their mother around the clock. Puppies also spend a great amount of time interacting with littermates. Thus, the interaction with the mother is not only important for the survival of the offspring, but also provides the foundation for the subsequent social behavior of dogs. While basic social responses are innate, the early experience of a puppy with its mother and littermates refines and develops basic responses, including the appropriate use of submissive gestures and the behavior of a subordinate. A pup's aggressive behavior is also shaped by interacting with other puppies.

If puppies are only exposed to their littermates and have little contact with people, they tend to develop strong attachments only to dogs. They may never form attachments to people, and may even exhibit fear and escape responses toward them. If a puppy is removed at about 3 weeks of age from its littermates and then only exposed to people, it may be primarily attached to people and show abnormal social responses toward other dogs. Rather than resolving conflicts by threat and submission, for example, it may readily fight with other dogs. Normally, puppies are exposed to dogs and people during the socialization period, and as adults react normally to both. It is usually recommended that a puppy be taken from the litter to its permanent home at the age of 6 to 8 weeks, thus allowing for adequate socialization first to its littermates, and then to its new human companions.

PROBLEMS WITH MATERNAL BEHAVIOR

Given the major physiologic changes that occur in a bitch at the time of birth, let alone the overwhelming presence of demanding puppies, it is surprising that female dogs remain as calm as they do during this time. One of the interesting hypotheses about why rat mothers remain calm after birth may apply to dogs as well. It has been found that stressful physical stimuli produce a smaller adrenocortical response in lactating females (Stern and Levine, 1972). The idea is that rather than responding to threatening stimuli with general systemic activation, the mother is induced to remain calm physiologically and continues to provide for the young. In the wild this would put the mother's own safety in jeopardy, but the trade-off is that the mechanism does prevent disruption of the care of her offspring at a critical time.

Anxiety, Nervousness, Cannibalism

A major problem arises with maternal care when the mother appears to become excessively nervous. This emotional behavior may lead to the attack of her own newborn. If she consumes any part of the dead newborn, the behavior is termed cannibalism. Killing and cannibalism of newborn has been observed not only in dogs and cats, but in wild animals as well, and may represent a normal aspect of the reproductive process. Postparturient female hamsters, for example, almost invariably kill and consume some of their offspring. It has been found that female hamsters adjust their litter sizes in accordance with the environmental conditions and food supply prevailing at the time of parturition (Day and Galef, 1977). Thus, pup cannibalism can be regarded as a normal aspect of maternal behavior in some species.

There is evidence in other species that cannibalism is brought on by the nutritional needs of the mother. In squirrel monkeys, abortion and cannibalism occurred with females on a low-protein diet but were never observed in females on a high-protein diet (Manocha, 1976).

Whether true or not, cannibalism in dogs is often considered to be related to the immaturity of the mother, lack of maternal experience, illness of the newborn, hyperemotionality, and environmental disturbances. There may be circumstances when cannibalism can be considered normal. Sickly puppies might harbor disease organisms that could be transmitted to littermates. If a mother attacks and consumes the sickly pup she loses one but saves the rest of the litter. The trigger for such cannibalism may be cold or inactive pups. To make this system work, the mother's cannibalism must be triggered by any slight abnormality in the pups, such that the sick pup is disposed of before it infects its littermates. Since mother dogs do not actually diagnose a disease but operate on signs that are correlated with disease, her cannibalistic behavior could be triggered by environmental disturbances unrelated to actual disease.

There may also be hormonal factors that incite a mother to attack her infants. The placenta produces appreciable amounts of progesterone during pregnancy, but the level falls at parturition with the detachment and expulsion of the placenta. Since progesterone has calming properties, the decline in progesterone may precipitate irritability and aggression toward the young, especially since other disturbances are occurring at the same time.

Maternal Indifference

Maternal indifference may be a result of domestication as explained at the beginning of this section. It might also reflect a hormonal defect in the endocrine changes that help bring about maternal behavior. We know that the strong attachment of a bitch for newborn puppies, as well as specific behavioral "housekeeping" tasks, are facilitated by hormonal changes at the time of birth. The specific hormonal conditions that produce maternal responsiveness are not properly understood even in laboratory animals. An alteration from normal hormonal secretions could underlie a deficiency in the induction of optimal maternal care.

The various stimuli from newborn puppies, i.e., sight, smell, and sound, also help activate and maintain maternal behavior. One study on rats, for example, showed that a single rat pup does not provide enough stimulation to the mother to maintain satisfactory maternal behavior and lactation even for the one newborn itself (Leigh and Hofer, 1973). If this applies to dogs, the work suggests that the bitch with a litter of only one pup should be carefully observed so that compensative measures can be taken if she does not provide adequate maternal care.

Pseudopregnancy

Before this behavioral syndrome was understood, pseudopregnancy in dogs was considered an abnormality. This is probably because, of all domestic female animals, only the bitch has been observed to show behavioral signs of pseudopregnancy. At one time the dog was considered the animal model of the classic human behavioral syndrome characterized by a barren woman whose obsession with wanting a child caused her to experience the physiological and behavioral changes associated with normal pregnancy. We now know that pseudopregnancy in barren female wolves, whose siblings are pregnant, is highly functional in a wolf pack where there is pairing between just the dominant male and one female. Other females, who are usually aunts of the pups, can serve as nursemaids, and in this way they contribute to the survival of offspring related to them and indirectly enhance their reproductive output (Voith, 1980c). Female dogs retain this capacity to varying degrees, so the behavior we see should be considered normal, albeit usually undesirable.

In dogs the syndrome starts prior to parturition with the bitch showing enlargement of the mammary area and abdomen. The mammary glands may develop to the point of secreting milk. Many dog owners, not knowing whether the dog was mated or not, fully expect that the bitch is pregnant. Pseudopregnancy usually subsides within the last two weeks of the expected parturition. In some bitches, however, the pseudopregnancy continues into parturition with the female displaying behavioral signs of impending parturition, including abdominal contractions and straining. Soon afterward the dog may collect a few stuffed toys and treat them as newborn puppies, licking and hovering over them as if to nurse them. This variant of maternal behavior may continue for as long as a normal lactation, ending at the expected time of weaning, when the bitch abandons the adopted toys and her behavior returns to normal.

Pseudopregnancy does not occur in spayed bitches. The syndrome appears to be brought on by the secretion of progesterone by the ovarian corpus luteum which forms after ovulation occurs. However, spaying a dog in the postparturient phase of the condition does not immediately eliminate the behavior. The hormone that may be at least partially responsible for maintaining maternal behavior in the postparturient period is prolactin. The blood levels of two other hormones in female dogs during pseudopregnancy, estrogen and progesterone, are no different than that of female dogs not displaying this behavior (Smith and McDonald, 1974).

Administration of sex steroids will often suppress the signs of pseudopregnancy by negative feedback inhibition of the hypothalamic-pituitary axis. A common treatment is megestrol acetate for a week or longer (2 mg/kg once daily for 8 days). Of course, spaying a female dog will prevent the condition from occurring in the future.

Chapter 12

Aggressive Behavior in Cats

Aggression is a part of every animal's behavioral repertoire. In contrast to dogs, aggressive behavior toward people is not often a problem in cats. This is a reflection of the social behavior of cats; they do not interact with us in a dominant-subordinate manner and, therefore, do not attempt to gain dominance by aggressive responses. Cats can be fearful of people, but fear-related aggressive behavior is not a major problem in cats since they can easily hide or escape from people that evoke fear. We also do not tend to force cats to interact with people.

Tomcats are drawn into fights with each other more frequently than gonadally intact male dogs. This is because tomcats do not have the innate social mechanisms of settling agonistic encounters by gestures and postures that communicate and establish dominance or subordination.

These differences in the types of aggressive behavior seen clinically in dogs and cats underscore the importance of understanding the differences between the species in treating problem behavior. A modification of Moyer's (1968) classification of types of aggressive behavior is as useful in cats as it was in understanding canine aggressive behavior. The types of aggression seen in cats are territorial, intermale, fear-related, pain-induced, and predatory. Of these, intermale and fear-induced aggressive behavior merit the most extended discussion.

INTERMALE FIGHTING

Male cats, both tomcats and even some castrated males, have a tendency to fight with other males. This behavior is sexually dimorphic and is a manifestation of the early masculinization process resulting from the exposure of males to perinatal gonadal androgen secretion (see Chapter 23). This behavior is clearly facilitated or enhanced by postpubertal androgen secretion, and it is the one type of aggression that is largely prevented or eliminated by castration, which, of course, is routinely practiced with cats. A clinical survey indicates that when cats are castrated in adulthood, approximately 90% of them stop fighting either immediately after the operation or gradually over a period of weeks or months (see Chapter 23). A survey of prepubertally castrated males reveals that fighting is not uncommon when these cats reach adulthood and that the percentage of prepubertally castrated cats that engage in serious fighting is about the same as the percentage of male cats that continue to fight after they have been castrated in adulthood.

Castrated males that continue to engage in fighting can be given a long-acting progestin such as injectable medroxyprogesterone or oral megestrol acetate, which in most instances will reduce or eliminate this fighting (see Chapter 23). Progestins appear to extend the effects on fighting that one would expect from castration. As an example, Jeremy (Case 12-1) is a male cat that was given progestin treatment for intermale fighting.

FEAR-RELATED AGGRESSION

Cats, like other species of animals, display fear-induced aggressive behavior when they wish to escape but realize that escape is impossible. This is the most frequent reason that people are bitten by cats, and it is the type of aggression associated with the classic black-cat Halloween posture. If a cat is backed into a corner it will attack. The encounter lasts only long enough for the assailant to be driven away. A cat will not pursue to the point of drawing the assailant into a fight.

We can manage this behavior by simply not exposing cats to fear-inducing stimuli. If the aggressive behavior is toward a person, some cats will gradually adapt to the person and the problem is self-correcting. There is usually no need for a desensitization program as there is in cases of fear-biting in dogs.

The use of tranquilizers, especially diazepam (Valium) may suppress fear-induced reactions and facilitate adaptation. Once it is determined

CASE 12–1

History. Jeremy, a prepubertally castrated male, gets into serious fights with other cats in the neighborhood about every two weeks during the breeding season.

Diagnosis. Intermale Fighting and Roaming

General Evaluation. Jeremy will be treated with a long-acting progestin that may reduce his fighting and roaming in the neighborhood.

SPECIFIC INSTRUCTIONS

1. Jeremy is being given an injection of medroxyprogesterone acetate (Depo-Provera). If this works, it will last 1 to 2 months.
2. If the treatment is not successful, return for an alternate drug treatment with an oral progestin.
3. To help diminish the roaming, tell friends who see Jeremy at their homes to spray him with a garden hose.

that a drug has a favorable effect, it is best to continue exposing the problem animal to the stimuli that evoked the aggressive behavior to desensitize it to the fear-evoking stimulus. The level of the tranquilizer should then be gradually reduced over a period of weeks in an effort to complete the desensitization process (see Chapter 24).

OTHER TYPES OF AGGRESSION

Territorial Aggression

This is a behavior that keeps asocial animals spaced apart. Cats often prefer to avoid each other but when one cat intrudes into another's territory, or if they are forced together, a fight is likely to occur. An adult cat, male or female, is unlikely to accept new cats, including kittens. Although a liaison between adult cats can develop with time, they may remain permanently intolerant of each other. In these cases, a satisfactory resolution is not always possible. For example, when Wuzzy (Case 12–2) became aggressive when a new kitten joined the household, the clients ultimately needed to make the choice as to whether the level of aggression was tolerable.

Pain-induced Aggression

Painful stimulation is a reliable elicitor of aggressive behavior. Punishing a cat by slapping or hitting it may be followed by scratching or biting by the cat as a direct response to the pain. Children are occasionally attacked by cats when they grab a cat's ear, tail, or hair. This contrasts with the effects of punishment in dogs, where the pain or censure frequently evokes a subordinate or submissive gesture. Since cats are asocial,

CASE 12–2

History. Wuzzy's inability to accept a new kitten is creating problems in the Morrison household. Wuzzy has been a wonderful cat for twelve years, but she has acted aggressive with the new kitten and has been aggressive and withdrawn from the owners since the kitten arrived.

Diagnosis. Territorial Aggression

General Evaluation. Wuzzy is asocial, like any cat, and he was never exposed to other cats. It is unlikely that Wuzzy will accept the kitten as a playmate. The most you can hope for is peaceful coexistence. Your main concern should be to induce Wuzzy to return to his previous behavior toward people.

SPECIFIC INSTRUCTIONS

1. Option 1: Try keeping the cats together for two more weeks. It may appear that Wuzzy will not return to his former behavior while you have the kitten, but you may want to be sure.
2. Option 2: Board the kitten for one week and observe Wuzzy for a return to normal behavior. If Wuzzy regains his past behavior, consider finding a new home for the kitten.

they rarely take the subordinate role and instead automatically respond to punishment by biting or scratching.

Predatory Aggression

The stalking and killing of rodents and birds by cats may be thought of as aggression, but it is so different from the primary types of aggression that it will be dealt with as a separate topic in Chapter 17.

The Petting and Biting Syndrome

Male cats, both castrates and intact animals, occasionally bite or severely scratch a person after they have been handled and petted for several minutes. The cat initially seems to enjoy the handling and petting, which then apparently reaches some threshold, and the cat suddenly turns and attacks. Progestin treatment, the same as with intermale aggression, is often effective in treating this problem.

Chapter 13

Inappropriate Urination and Defecation in Cats

A stereotype of the cat is that it is fastidious in personal hygiene. No other domestic animal digs a hole for elimination and then covers it. Most cats seem so fixed in this behavior that we hardly need to give a second thought to housebreaking most cats. People have speculated from time to time about the evolution of this unique burying behavior. Certainly it aids in helping a cat or a mother's kittens avoid ingesting parasite eggs or intestinal pathogens through contact with uncovered feces. Another possibility is that covering feces reduces the evidence of a cat's presence to the rodents it preys on in its territory. Still another possibility is that burying the feces keeps tapeworm segments (and eggs) out of the reach of rodents who are intermediate hosts of tapeworms. Cats can reinfect themselves by eating rodents who have encysted tapeworms in their muscle tissue.

The burying behavior so characteristic of cats makes them particularly nice house pets, even for apartment dwellers. An owner needs only to provide a litter box that is regularly cleaned. Some people have disguised the litter box so that it is hardly identifiable. One cat owner built a kitchen cabinet large enough for a litter box and a deodorizer bottle. Cabinet doors could be opened in the front for easy removal and cleaning. Cats could enter the cabinet at will by a small opening on the side. Visitors did not notice the litter box. The commercial versions of this idea include a plastic enclosure which disguises a litter box and also prevents access of dogs to feces in the litter.

FACTORS AFFECTING ELIMINATIVE BEHAVIOR

As well-known as cats are for their fastidious eliminative habits, it is still necessary to address behavioral problems dealing with inappropriate elimination. Since urine spraying and marking problems require a completely different type of therapeutic approach, they are dealt with in Chapter 14. In this chapter we will consider the cat that for some reason starts to defecate and urinate in inappropriate areas of the house rather than outdoors or in a litter box. To approach this problem requires some understanding of the cat's natural eliminative tendencies, as well as the reason why a cat may find the litter box aversive. We must also evaluate whether certain diseases may have precipitated this problem.

Some cat owners confuse urine spraying with inappropriate urination. Inappropriate urination involves the usual squatting posture, but differs from normal urination in that a variety of areas, all unacceptable to the owner, are used rather than a litter box or spot outside. In urine spraying, the targeted area is 1 to 2 feet off the ground level on vertical objects. The approach for treating spraying is different than for inappropriate urination, but there are instances where cats both urine spray and urinate in inappropriate areas.

Inappropriate urination can result from urinary infections. For example, cystitis causes a tendency for frequent urination that may result in the animal not being able to make it to the litter box each time. If there are signs of a urinary disorder, a urinalysis should be conducted.

Other factors to consider are senility, which may lead to increased frequency of urination, weakened urinary sphincter control, or mild arthritis that makes movement more difficult. Just an aversion to going outdoors in inclement weather may predispose a cat to inappropriate elimination.

The elimination pattern of cats protects them by reducing exposure of the animals to the eggs of intestinal parasites either directly or through intermediate hosts. In the course of evolution, this behavior has been maintained genetically because of the threat of disease. As we have domesticated cats and controlled these diseases, we have removed the natural selection pressures that maintained the innate basis of eliminative behavior. Cats that are not very sanitary in their eliminative behavior live and reproduce as much as those that are fastidious. This contributes to variability in the genetic mechanism controlling eliminative behavior, and explains why some cats cannot be trained to be as sanitary as we would like.

THERAPEUTIC APPROACHES TO PROBLEM ELIMINATION

Correcting Aversions to the Litter Box

Inappropriate urination and defecation can reflect an aversion to the litter box or an attraction to another area. Understanding the effects of aversions may be as simple as noticing that the litter box is not being cleaned frequently enough. Daily cleaning may be sufficient for a single cat. A person may not realize that when he gets one or two additional cats, the litter box should be cleaned proportionately more often. Cats may avoid a litter box that is infrequently cleaned.

Aversions may result from changing litter material or introducing an additive, such as chlorophyll. Apparently many cats find chlorophyll and other deodorizing additives aversive. The problem may be solved by the owners shifting to unadulterated litter. Some cats may not even like claylike litter, and it may be necessary to use plain sand.

An aversion to the litter box can be caused if people take advantage of the animal's vulnerability while defecating or urinating to administer pills or injections. This will teach the cat to avoid the litter box. If a family has recently adopted a kitten, it might be necessary to discuss care of cats and point out that adults and children should not disturb a cat when it is using the litter box.

The smell of an area that has been previously soiled is a source of attraction for cats. When reestablishing a litter box habit, the soiled areas of the house should be thoroughly cleaned. Carbonated soda water is sometimes useful but ammonia, because of its similarity to urine, should not be used.

Reestablishing Litter Box Habits

The most recommended procedure is to confine the cat to an area where there is a high probability of it using the litter box. Frequently the best place is the bathroom, because the tile floor is not an acceptable toilet area. As the cat begins to regularly use the litter box, it can gradually be allowed access to other parts of the house. Thus, the cat might have free run of the house for a couple of hours when it can be watched fairly closely. Alternatively, it could just be allowed into an adjacent room or hallway. Generally a cat soon earns free run of the house with the litter box being located in the most convenient area for the owner. This process was successful in treating Bozo (Case 13–1).

A procedure for a carpeted room would be to cover the carpet with

CASE 13–1

History. Although Bozo, a castrated male, used to use his litter box unfailingly, a month ago he began avoiding it while it was unchanged for a few days during Mr. Harriman's absence.

Diagnosis. Inappropriate Elimination

General Evaluation. The therapeutic approach is to gradually lead Bozo back into his former toilet habits.

SPECIFIC INSTRUCTIONS

1. Encourage Bozo to spend more time outdoors.
2. When Bozo is inside, confine him to a small bathroom where a clean litter box is available.
3. As Bozo shows success in using the litter box, extend his freedom. Allow him some access to other parts of the house for brief periods while you watch him. Continue to keep him in a confined area after he eats, or at those times of day when he regularly eliminates.

a plastic sheet and place the litter pan on the carpet. As the litter pan is regularly used, the plastic can be gradually trimmed away.

Rupert's case illustrates a more complicated approach to the correction of toilet habits. Rupert had decided that the bathroom carpet was more to his liking than the litter box which was usually kept in the bathroom. It was suggested that the owners temporarily remove the bathroom carpet, and since the carpet was ruined anyway, cover the bottom of the litter box with the old carpet. The attractiveness of the carpet was revealed when Rupert immediately began using the litter box. After each cleaning of the box the owners were instructed to sprinkle commercial litter material on the carpet, adding a bit more each day until the carpet was uniformly covered with about one half of an inch of litter (it took about one week to reach this point). The cat continued to use the litter box, so the next phase was for the owners to trim away about one inch of the material from the perimeter of the carpet after each cleaning until the carpet was completely gone.

Toilet Training

Not only can cats be fastidious in toilet habits, but they can be quite adaptable to modern plumbing amenities. Aside from being a conversation piece, having a cat use a toilet as depicted in Figure 13–1 not only saves litter material but eliminates the odor problem. To begin toilet training, a normal litter box is kept in the bathroom until the cat is regularly using it in this location. Then the litter box is removed and a different "litter box" is made from the toilet seat. A cardboard rim should be cut in the shape of the toilet seat and this used to support a sheet of clear plastic to the underside of the toilet seat (with wires). Then, when litter material is placed in the toilet seat, the rim of the seat forms the edge of the litter box and the plastic sheet contains the litter. With plenty of litter material in the toilet seat, cats readily use the new litter site which, of course, temporarily removes the toilet from human use.

Once the cat has become accustomed to using the toilet seat with litter, the amount of litter is reduced by half. Meanwhile the cat has learned to stand with its feet on the toilet seat rather than in the litter. More litter is gradually removed and holes are made in the plastic to allow urine to drip through. Whenever the cat stops using the toilet seat because litter has been removed too rapidly, the owners will have to return to an earlier stage in the training procedure. Finally, all of the litter and the plastic sheet are removed, and the cat is regularly allowed to use the toilet with no intervention or assistance from the owner.

It is expected that a cat will slip off the toilet seat at least once and fall into the toilet bowl. Such an accident may result in the cat requiring retraining to the toilet seat by placing litter again on top of a sheet of plastic.

A properly toilet trained cat is such an appealing prospect that the

Fig. 13–1. This cat was toilet trained by first making the toilet seat into a litter pan (litter held by plastic sheet over bottom of toilet seat) and gradually removing litter material over several days (From Hart, 1975b; reprinted with permission, Veterinary Practice Publishing Co.).

idea has naturally attracted the attention of commercial interests. One company has manufactured preformed plastic trays that fit over the toilet seat. This makes the technical approach to toilet training easier and allows people to use the toilet during training by simply lifting the tray off.

Chapter 14

Urine Spraying and Urine Marking in Cats

Objectionable urine spraying and urine marking constitute the most frequent behavioral problems in cats. The behavior is normal for gonadally intact adult males, and prepubertal castration usually prevents the occurrence of the behavior. However, castrated males and females that are either spayed or gonadally intact may also spray or urine mark. Recent information reveals that about 10% of prepubertally castrated male cats and 5% of prepubertally spayed female cats take up spraying on a frequent basis as adults (Hart and Cooper, 1984).

The behavioral act of spraying is most often initiated by the cat smelling the target area, which is usually a vertical surface a foot or so above ground level. The cat then turns around and directs a stream of urine toward the investigated area. Males and females spray using basically the same posture. Frequent targets for spraying inside are interior walls, draperies, furniture, bookcases, and kitchen appliances. Frequent outdoor targets are trees, bushes, doorsteps, and automobile tires.

Most tomcats urine mark their territory by spraying, but females and castrated males may urine mark using the squatting posture, and in places that are not normal toilet areas. Frequent targets for this type of urine marking are specific places on a carpet, the owner's bed, or clothes of a family member. For the purposes of treatment, the two forms of urine marking will be considered as one type of problem behavior.

Urine spraying and marking are basically innate behavioral patterns that can prove difficult to control by punishment or management. Behavioral treatment is useful for urine spraying or marking that occurs in only one or two specific areas. Progestin treatment is more likely to be effective for more widespread spraying but depends on the sex and the number of cats in the home. Although more extreme, a neurosurgical approach (olfactory tractotomy) may be desired by some cat owners when progestin therapy is ineffective and euthanasia is the only other alternative. The therapeutic measures will be discussed after some background material is reviewed.

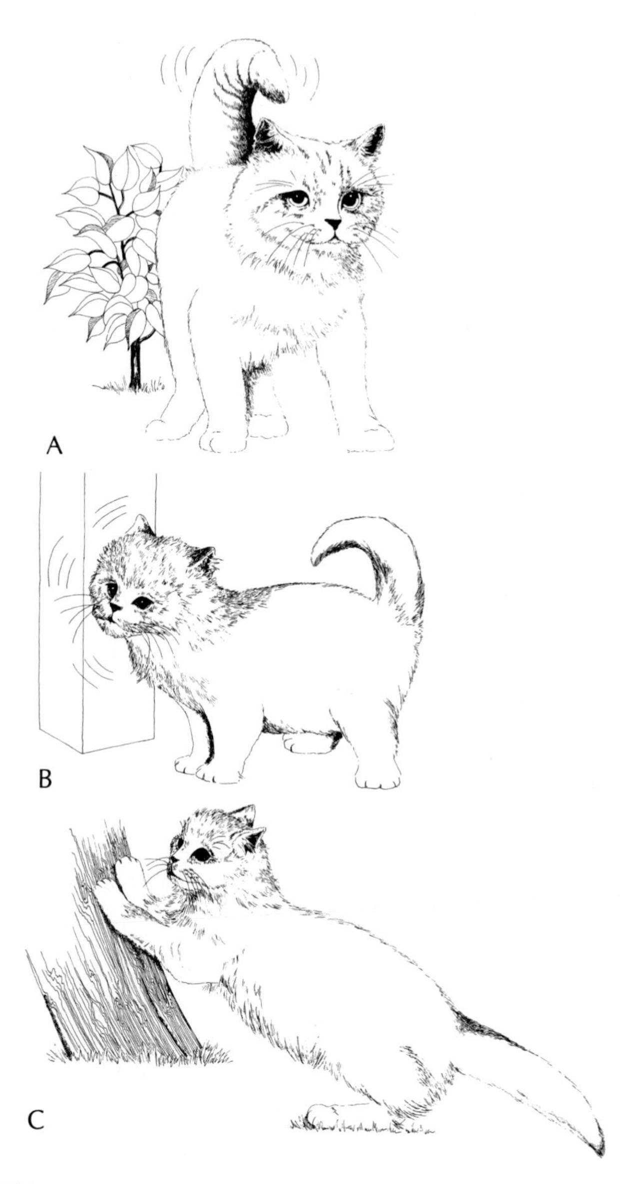

Fig. 14–1. Urine spraying *(A)* is one of three types of scent marking by cats. In addition to urine spraying, this illustration shows scent marking by a cat rubbing secretions from glands at the corners of the lips *(B)* and from glands located in the feet when the cat scratches something *(C)*.

DETERMINANTS OF SPRAYING BEHAVIOR

Cats have three readily identifiable scent marking behaviors (Figure 14–1 *A–C*), one of which is urine spraying. Spraying is a type of territorial mark that tomcats primarily engage in. The marking familiarizes the male with his territory and home range. His own urine odor probably makes him more self-assured and comfortable and also communicates his presence to other cats in surrounding areas. An important cause of spraying is an increase in anxiety or nervousness. Cats probably do not intentionally mark the boundaries of their territories (as one would put up a fence), but rather they mark boundary areas because this is where anxiety-provoking encounters occur with other cats. Spraying is also undoubtedly useful during the breeding season in attracting sexually receptive females to his vicinity. Spraying does not result from cognitive processes, such as a scheme to express resentment at the owners. Its occurrence has increased through evolutionary or adaptive responses favoring the reproductive success of the cat's feral ancestor. Spraying frequency by males usually increases during the feline breeding season. This may result from a seasonal increase in testosterone secretion, social interactions, and a higher level of sexual awareness.

Intact male cats vary in the degree to which they engage in spraying. Spraying may be influenced by the number of cats housed in the area, or by changes in the environment. For example, during a period of adjustment to a new house after a move, there may be a transient period of spraying activity. A nonspraying cat may start spraying when new cats enter the neighborhood or the household.

Intact male cats have a much higher tendency to spray than male cats that have been castrated prepubertally or in adulthood. However, even males that were castrated at four months of age may start spraying if anxiety-provoking circumstances become intense. For example, during the feline breeding season there is a general increase in social interactions, such as roaming, fighting, and spraying by other male cats. Introducing new cats into a neutered cat's home range or territory may also evoke spraying.

In nature, spraying is a normal behavior for some intact females during the breeding season. By depositing sexual attractants through her urine on vertical objects in her home range or territory, she attracts male cats to her vicinity.

One factor that can stimulate gonadectomized cats to spray is new contact with other cats. Contact with other cats can lead to repeated agonistic encounters, and these emotional disturbances can stimulate cats to spray. As it turns out, the probability of female cats spraying is not affected by the presence of other cats, but with castrated male cats, the presence of female housemates increases the probability of spraying much more than the presence of male housemates (Figure 14–2).

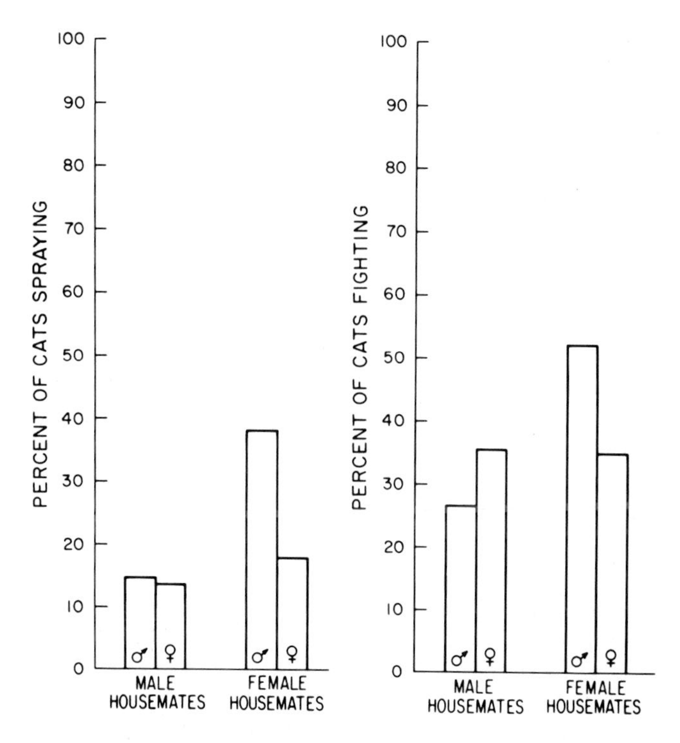

Fig. 14–2. The likelihood of male cats engaging in urine spraying is much greater in households where the male cats have female littermates than in those with male littermates. This graph illustrates the prevalence of occasional or frequent spraying or fighting in male and female cats in households with male and female housemates. (From Hart and Cooper, 1984; reprinted with permission, American Veterinary Medical Association.)

THERAPEUTIC APPROACHES

Diagnostic Problems

Since cat owners differ in their familiarity with spraying, one of the first tasks when presented with a potential sprayer is to determine whether the problem is spraying or inappropriate urination. Treatment of the latter is discussed in Chapter 13. For any cat with a presenting complaint of spraying, it is important to rule out urinary infections by the medical history or, if necessary, by urinalysis. In recurrent instances of spraying, a urinalysis is indicated. Urinary cystitis seems to increase the tendency of a male cat to spray as well as urinate more frequently in the nonspraying fashion.

When treating urine spraying and urine marking it is a good idea to explore the reasons why the animal started spraying and continues to spray. Did spraying begin with the onset of the breeding season? Did the owners recently move? Is the cat especially nervous or anxious recently? Are there new cats in the household? Some factors that provoke spraying may be transient, such as the onset of the breeding season or the owners moving to a new house, and therefore, once the spraying has been reduced or eliminated the problem may be solved.

In multi-cat households, a problem in diagnosing urine spraying is to determine which of the cats is actually doing the spraying. Although one cat may be observed while spraying, it is possible that another resident cat also sprays. The owner could separate the cats; however, this changes the environment enough so that the cat doing the spraying may temporarily stop. To mark the spray, sodium fluorescein dye, of the type used to reveal corneal ulcers, may be given to the cat suspected of urine spraying. If the dye is given orally, or injected subcutaneously, it is readily excreted into the urine. Urine-soiled spots retain fluorescence for at least 24 hours, and during subsequent days the fluorescence gradually deteriorates. The dye appears in the urine within two hours of oral or subcutaneous administration. It is water soluble and, when diluted in urine, it does not discolor fabrics.

To identify the spraying cat, the following procedures are suggested for using sodium fluorescein as a marker. The cat considered the most likely source of spraying should be given an injection subcutaneously of 0.3 ml of sodium fluorescein (10%, equivalent to 100 mg/ml) in the late afternoon. The urine deposited that night and the next morning will then be labeled (Fig. 14–3). Alternatively, the client can administer the dye to the cat orally at home by giving 0.5 ml of the solution, or 6 strips

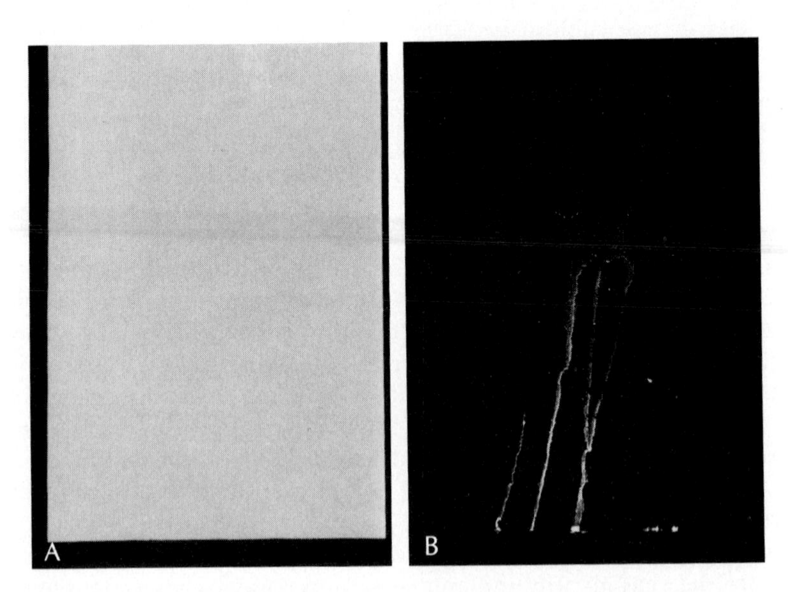

Fig. 14–3. Treating cats with sodium fluorescein does not discolor urine sufficiently to be visible by conventional lighting *(A)* but makes urine marks quite obvious with ultraviolet (Wood's lamp) illumination *(B)*. (From Hart and Leedy, 1982; reprinted with permission, American Veterinary Medical Association.)

of ophthalmic test paper inserted in gelatin capsules. After subcutaneous or oral administration, the client can be loaned an ultraviolet (Wood's) lamp, and instructed in how to scan the house for urine spots while the house is darkened. The possible culprits for urine staining can be treated at 2-day intervals until the owner is fully satisfied as to the urine source (Hart and Leedy, 1982).

Castration

Tomcat urine has a pungent odor and is distinctive from that of females or castrated males. Apparently androgens produced by the gonads cause the odor, but the nature or source of the odor is not known. Castration readily eliminates this odor within a few days after castration (Bland, 1979). Although the odor of male cat urine is made less offensive by castration, dealing with spraying behavior is a more complex issue. Castrating tomcats after puberty is quite effective in eliminating or markedly reducing spraying, even after spraying has begun. In a clinical survey conducted several years ago, it was noted that about 80% of adult males castrated because of a spraying problem underwent a rapid decline in the behavior, with an additional 10% experiencing a more gradual decline (Hart and Barrett, 1973). From these data one could expect that approximately 10% of male cats castrated because of a spraying problem persist in the behavior indefinitely.

Persistence in spraying following castration is not caused by residual amounts of testosterone, since within 8 to 16 hours after castration, the concentration of testosterone in the blood is reduced to castrate levels (Hart, 1979d). Persistent urine spraying or urine marking in male cats that have been castrated as adults is treated the same as that for prepubertally castrated male cats discussed below.

Male cats are commonly castrated before puberty, and veterinarians and cat owners commonly assume that prepubertal castration is more effective than postpubertal castration in preventing objectionable urine spraying, fighting, and roaming. Male or female cats, however, neutered at 6 months of age, may begin spraying at three or four years of age. Several questions regarding the relationship between castration and urine spraying inevitably come to mind. Is prepubertal castration actually more effective than postpubertal castration in eliminating urine spraying? Are neutered males more likely to engage in spraying than females? Is the age of the cat at the time of prepubertal gonadectomy related to the likelihood of spraying? With female cats there is the issue of whether a female laying in the uterus adjacent to one or two males may have been partially androgenized and is thus more predisposed to engage in male behavior, including urine spraying, than females from all female litters. Prenatal androgenization of behavior such as this has been documented with regard to sexual and aggressive behavior in female rats and mice (Clemens, 1974; vom Saal, 1981).

These questions were addressed in a recent clinical survey of 134 male and 152 female cats (Hart and Cooper, 1984). No relationship was found between the age of male cats at the time of prepubertal castration (which ranged from 6 to 10 months) and the likelihood of spraying. There was also no relationship between the age of ovariohysterectomy in female cats and the incidence of spraying. The incidence of frequent spraying by prepubertally spayed females was 5%. The spraying tendency in females coming from litters in which all the other littermates were male, and in which androgenization was likely, was no greater than in females coming from all female litters.

The incidence of frequent urine spraying by prepubertally castrated male cats in the survey was close to 10%. This is the same proportion of male cats that are castrated in adulthood for problem spraying and continue to spray. Thus, prepubertal castration is not any more likely to be effective in preventing objectionable spraying than postpubertal castration is in eliminating the behavior once it has started. Cat owners who wish to allow their males to grow the larger head, heavier jowls, and general morphology characteristic of tomcats do not increase their risk of having a spraying problem by delaying castration into adulthood.

Behavioral Management

Remember that in dealing with spraying one is attempting to alter a cat's normal response to certain social or environmental situations. Behavioral approaches to control spraying are usually disappointing.

Cat owners often try punishment, such as yelling at the cat or throwing something at it when it is caught in the act of spraying. In some instances this approach has been effective, and the cats, although still spraying regularly outside the house, do stop spraying inside. Most cat owners understand the futility of bringing a cat to the urine-soiled spot and either pointing it out or rubbing its nose in it. Such punishment only makes the animals wary of the owner.

Remote punishment, using upside-down mousetraps near the soiled areas (see Chapter 22), is most effective. Ambushing a cat with a squirt gun or water sprayer when it is beginning to spray is another technique. It is important that remote punishment be delivered without the cat knowing that the owner is involved in the punishment process. In this way the animal makes the association between the target areas and the punishment rather than between the owner and the punishment. Remote punishment is only useful if a cat is spraying very few objects. Obviously one cannot have an entire house booby trapped with upside-down mousetraps.

If a cat only sprays in one or two spots, another technique would be to feed the animal at these spots during its regular mealtime. Cats are unlikely to eliminate in the same spots where they eat.

When spraying is occurring at only one or two sites, remote punishment is appropriate for the first round of treatment, and then if that is unsuccessful, additional steps may be necessary. In Case 14–1, the client did not know which of her three cats was spraying. During treatment she administered remote punishment and also identified the spraying cat so that additional action, if necessary, could be taken.

Progestin Therapy

Long-acting progestins are widely recognized as somewhat effective in correcting spraying. The preparations that are commercially available in the United States are medroxyprogesterone and megestrol acetate (see Chapter 23). One injection or an oral treatment series may permanently eliminate the behavior, especially if transient environmental factors originally evoked the behavior. If the cat is continually anxious or nervous or other factors are tending to stimulate the behavior, repeated injections may be necessary to keep the behavior suppressed. Among reported side effects is an occasional enlargement of the mammary glands. Owners whose animals have been treated frequently report two other side effects, lethargy or depression and an increase in appetite.

Clinical experience with progestin therapy indicates that problem spraying or urine marking is resolved for a month or longer in only about 30% of cats overall. However, the sex of the cat and the number of cats in the household are important considerations (Hart, 1980c). As Figure 14–4 illustrates, one can expect to see a response in males more frequently than females, and more in cats from single-cat households than in those from multi-cat homes. Thus, the prognosis for progestin

CASE 14–1

History. This client has three spayed female cats, and one of them sprays urine on two places: the refrigerator and stereo speakers. The client is interested in behavioral suggestions as well as possible drug treatment.

Diagnosis. Urine Spraying

General Evaluation. Remote punishment can provide some protection for the 2 spraying sites. Also, identifying the spraying cat will make it possible to take some other steps to curb the problem.

SPECIFIC INSTRUCTIONS

1. Identify the spraying cat by giving three fluorescein capsules to one cat, and then scanning the sprayed areas with an ultraviolet light for fluorescence. Treat each of your three cats with the capsules, one cat every two days, to confirm or eliminate them as problem animals.
2. Booby trap the two spraying targets, the refrigerator and the stereo speakers, by covering the area around them with inverted mousetraps.
3. If spraying continues or switches to a new location, bring in the cat which you have determined is spraying for treatment with oral or injectable progestin.

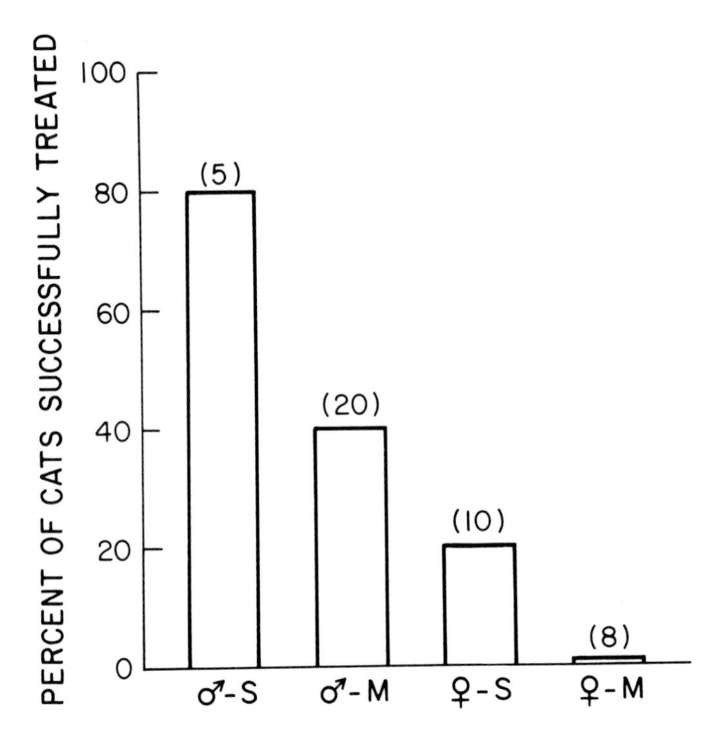

Fig. 14–4. Effectiveness of progestin treatment on urine spraying and urine marking in cats. A greater percentage of males than females respond and more cats from single than multi-cat households respond. When both gender and household environment are taken into account, the prognosis ranges from a high percentage (about 80%), to a very low percentage. (From Hart, 1980c; reprinted with permission, American Veterinary Medical Association.)

therapy is highly dependent on the sex of the cat and its home environment. Females from multi-cat environments warrant the poorest prognosis for progestin therapy, and males from single-cat homes have the most favorable prognosis. Figure 14–4 illustrates the prognosis one might make for the four different conditions. The best prognosis is for males in single-cat households and the worst is for females from multi-cat households.

Since medroxyprogesterone and megestrol acetate seem to be equally effective in the initial treatment of spraying, but megestrol acetate results in more depression and appetite stimulation than medroxyprogesterone (Hart, 1980c), we recommend that the injectable medroxyprogesterone be used in most cases for the initial treatment. Using an injectable drug also eliminates the necessity for the client to observe a complex dosage regimen. A dose of 10 to 20 mg/kg of medroxyprogesterone is recommended. Since the hair immediately over the site of injection occasionally changes pigmentation, subcutaneous injection into the inguinal region is recommended. The recommended initial oral dose of megestrol acetate is 5 mg/day, whether as an initial treatment or following unsuccessful medroxyprogesterone treatment. If effective, the treatment dose should be gradually reduced over intervals of 2 weeks to a dose of 5 mg once a week, and then terminated in about 2 to 6 months. If improvement is not seen in one week with the initial megestrol acetate treatment, additional megestrol acetate should not be given.

Olfactory Tractotomy

When progestin therapy is not effective, some cat owners may wish to consider more extreme measures to resolve a spraying or marking problem. An olfactory tractotomy has been used successfully to reduce urine spraying or urine marking in cats that did not respond to progestin treatment (Hart, 1981b; 1982; 1984). It is a simple neurosurgical procedure, carrying little surgical and recovery risk. The operation requires relatively little in the way of specialized instruments and neuroanatomic background. The evaluations of cats treated with this operation show that it is effective in most but not all cases that do not respond to progestins, and there are no major adverse side effects. The rationale for the operation is that spraying is usually initiated by the cat smelling the target area. Rendering the cat anosmic would then remove much of the motivation for spraying. The surgical procedure is basically that which has been commonly used in laboratory studies, where the olfactory tracts and caudal parts of the olfactory bulbs are approached dorsally through the frontal sinus. The surgical details of the operation are presented elsewhere (Hart, 1981b; 1985b).

The day following an olfactory tractotomy, cats may show little interest in food. One can usually stimulate a cat to start eating by placing meatlike baby food in its mouth or smearing the food on its lips. After they have eaten baby food, they will usually accept the usual semimoist or dry cat food. Following olfactory tractotomy, cats still appear to sniff at objects. Cat owners can be asked to conduct a "hidden food" test by placing meat under paper towels and determining if the cat can find the hidden food, and thus confirm the results of the operation.

The operation has proved successful in 50% of male cats and virtually all female cats operated on. Among males individual responses vary. One might find a cat that sprays inside the house while under postoperative confinement, but after it is allowed access to the outdoors, no further occurrences of spraying occur. No major undesirable behavioral side effects are evident following the tractotomy. An increase in affection is sometimes quite remarkable, however. In some cats an increase in appetite, or at least a willingness to consume a greater variety of food postoperatively, may be noted. Occasionally a cat may become a more finicky eater.

This neurosurgical approach represents the extreme of a continuum of therapeutic measures that are now available to eliminate the problems

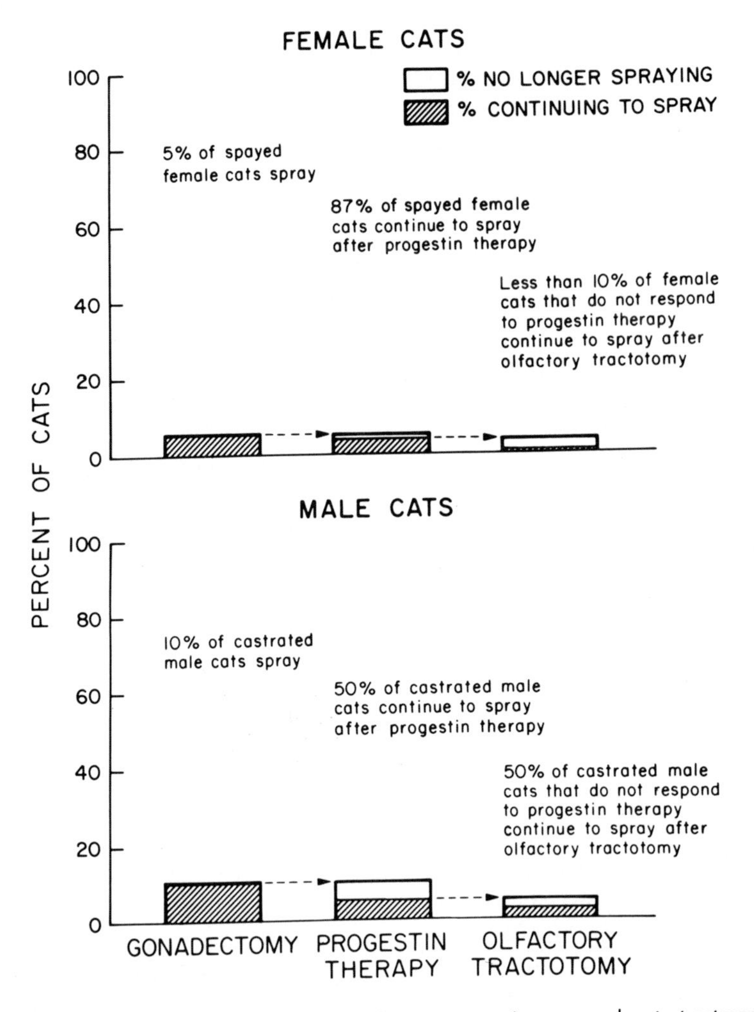

Fig. 14–5. Therapeutic results are represented using successive approaches to treatment of urine spraying in gonadectomized cats. (From Hart, 1985b; reprinted with permission, W.B. Saunders Co.)

of urine spraying and marking. The simplest, noninterventive measures are behavioral management techniques. As pointed out, these measures are often only temporarily effective. Progestin treatment is more effective but results depend on the cat's sex and the number of cats in the home. A neurosurgical approach could be considered for clients seeking additional alternatives. The advantages of the olfactory tractotomy procedure described here are that it requires little in the way of specialized equipment, carries minimal surgical risk, and is a relatively brief operation.

The successive approaches to treatment of urine spraying in male and female cats are graphically presented in Figure 14–5. The first treatment of choice is gonadectomy. Progestin therapy with either medroxyprogesterone acetate or megestrol acetate is then used to treat nonresponders. The neurosurgical approach of an olfactory tractotomy is the ultimate option.

Chapter 15

Training and Correction of Cat Scratching

Your typical mild-mannered cat can do practically as much damage to a house with its claws as a large dog can with its canine teeth. Scratching trees is an innate behavior of feral cats, and scratching our furniture is the equivalent for the more urbanized feline. This inherited behavioral tendency is so strong that some cats declawed early in life still routinely go through scratching motions. Like eliminative behavior, some cats scratch and the owners are not aware of it, while other cats make their owner's life perfectly miserable by scratching only the most valuable furniture. Declawing is a justifiable means of dealing with scratching problems, but some cat owners will not hear of it, and others will consider it only as a last resort.

In this chapter we will present background information on the factors that help attract a cat to scratching certain areas. This will be followed by practical suggestions for inducing kittens to scratch acceptable objects and therapeutic guidelines for an older cat with scratching problems. Providing advice on curing an old cat of scratching a valuable piece of furniture may win you some new friends and impress your clients. It is worth taking a conservative approach by offering some behavioral suggestions, and if these do not work, declawing is always available as a further option.

NATURAL SCRATCHING BEHAVIOR

Observing cats outdoors can teach us a lot about feline behavior indoors (Hart 1980a). Many outdoor cats have a scratching tree that is, of course, a prominent object in the environment. Since the cat repeatedly works over the same tree trunk with its claws, it becomes a personal territorial marker because the scratched appearance is readily visible to other cats that might venture by (Figure 15–1).

In the process of scratching trees, cats also rub secretions from glands in their feet onto the tree trunk; this way the scratched tree trunk gains a distinct olfactory character that can be recognized by other cats (see

Fig. 15–1. Typical scratched tree that is repeatedly scratched and serves as a visual and chemical territorial mark for cats.

Fig. 14–1, pg. 135). The chemical mark provides the resident cat with some familiarity with its territory, and the smell appears to attract the cat back to the same tree to restore both the visual and chemical mark.

Scratching also has the function of conditioning the claws. Claws are not sharpened as a knife blade would be sharpened, but are conditioned in that a frayed and worn outer claw is periodically pulled off by scratching, exposing a new, and very sharp claw already growing beneath. Worn claws that are removed by scratching may be seen at the base of scratching objects (Figure 15–2). Cats may also remove these outer claws with their teeth, particularly those on the back feet. Clawing is a natural, healthy behavioral requirement and is not something the animal does to "punish" or displease the owner.

The common declawing operation involves removal of claws from only the front feet. What are the behavioral effects of the operation? Most cats can still climb most rough-bark trees after the operation and use their back claws for defense. However, a cat's defense may be weakened because some cats are accustomed to fending off dogs by scratching with their front claws. Also a cat could find its life is endangered if it is unable to climb smooth-barked trees that it has routinely used for escaping before it was declawed. Surgical declawing is the only solution to some serious problems of furniture scratching. However, if a pet owner finds his cat fatally mauled after just being declawed, because of a sudden disruption of the cat's defense mechanisms, he may feel far worse than if behavioral approaches had been used to solve the scratching problem.

SCRATCHING-POST TRAINING

The occurrence of problem scratching in a house can best be approached from the standpoint of territorial marking rather than claw conditioning. Whether other cats have access to the house or not, cats

Fig. 15–2. Worn claws such as these that were removed by scratching can be found at the base of scratching tree or post.

have a strong, innate tendency to establish at least one territorial mark, preferably on a prominent vertical object of suitable texture. Frequently, the corner of a chair or couch that sticks out into the room is most visible and becomes chosen. As soon as the cat starts scratching a particular corner of a couch, it tends to persist there because the spot soon smells like a territorial mark that attracts the cat back again and again (Figure 15–3).

To induce a cat to choose an acceptable area to scratch in the first place, three principles should be kept in mind. (1) Once a cat starts scratching an object it tends to return to it; (2) prominent objects and areas are favored; and (3) the texture of a potential scratching area influences whether or not it is used. The importance of starting a cat to

Fig. 15–3. Furniture, especially prominent corners of couches or chairs, can become scratched territorial markers for cats inside, just as a scratching tree is outside.

scratch in the right place is emphasized in the first principle. A kitten can be directed to start on a scratching post before it starts scratching on the furniture. Obviously the post should be very prominent and kept in an area the kitten frequents until it is being used regularly. One should not wait until the animal is fully grown and capable of real damage before training it to a particular object. For a kitten it may be necessary to lay a scratching post horizontally to allow it to scratch and to develop an attachment to a particular post.

Whenever undesirable scratching occurs on furniture, the spot should be covered with plastic to prevent further scratching. A sturdy scratching post covered with appealing material should be put right in front of the same area, since this is obviously a desirable location for scratching. Temporarily moving the furniture aside is also a possibility.

Cats often develop propensities for scratching objects near their sleeping or resting areas, because they usually tend to scratch just after awakening. Scratching apparently serves as a form of stretching for the front limbs, as well as conditioning the cat's claws. Therefore, it is helpful to have at least one scratching post or board adjacent to where the animal sleeps.

Physical characteristics of the scratching object are also important. A flimsy scratching post that is easily tipped over is useless. Many experienced cat owners prefer a board attached to the wall rather than a freestanding post. A good dimension for a scratching board is 6 to 8 inches wide by 12 to 16 inches long. It should be adjustable in height as the cat grows. The best height is usually at least a foot off the floor so that the cat may comfortably rest its back feet while scratching.

Anyone who has lived with a cat scratching problem has probably noticed that cats tend to scratch some types of material more than others (Hart, 1980a). The material on the post should be something a cat likes. Most commercial posts are covered with carpet which, surprisingly, is too durable. As a covering becomes worn out and stringy, the cat likes it better because it can get a nice long drag, whereas with carpeting the scratching is jerky. With commercial posts that are covered with carpet, one might want to cover them with upholstery material that has longitudinally oriented threads like that shown in Figure 15–4. As the cat's smell is applied to the post material and the odor goes through the upholstery cloth, the cat will perhaps stay with the carpet long enough to get it somewhat stringy. A cat owner should not replace the carpet at the first sign of wear, since that is just when the cat begins to like it. Obviously cats do not replace an old scratched-up tree with a new one. As soon as the cat is habitually using a post, it may be moved (a few inches each day) to a spot preferred by the owner.

THERAPEUTIC APPROACH

Unless a cat has been declawed, it will usually require some object on which to condition its claws. While an outdoor cat will use a favored tree

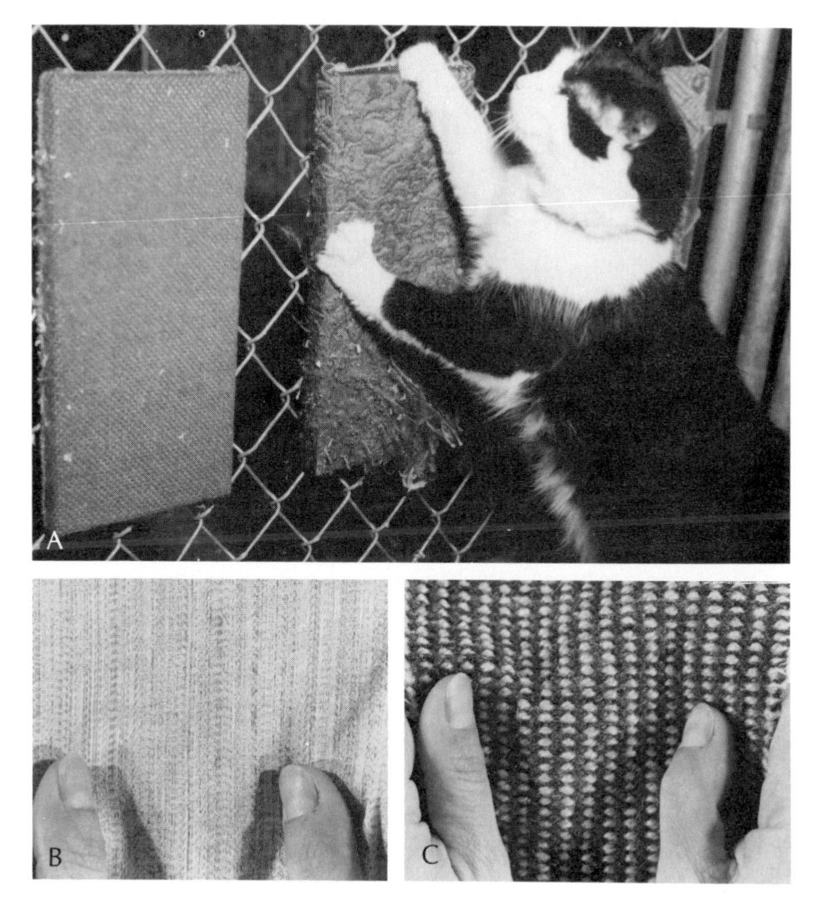

Fig. 15–4. Two-choice preference tests *(A)* reveal that cats prefer fabric with longitudinally oriented threads *(B)* for scratching rather than tightly woven knubby fabric *(C)*. (*B* and *C*, from Hart, 1980b; reprinted with permission, Veterinary Practice Publishing Co.)

or wood post, an indoor cat will need a scratching post or board. Of course there is no inherent reason why cats should prefer a scratching post to furniture. The scratching post label from the pet store does not impress the cat.

To train a cat to stop scratching, the furniture should be removed or moved and covered with some nonscratchable material, such as plastic or aluminum foil. Exactly where the scratched area of the furniture was, a scratching board or post should be placed. It should, of course, be covered with attractive scratching material. From this point on the process is the same as for training kittens. The post is gradually moved, inch by inch, to the edge of the room where it is not so obvious. This is the approach used to resolve a scratching problem with Tassle (Case 15–1).

In many instances damage to old furniture has been so severe that

CASE 15–1

History. The owner of Tassle, a 3-year-old black and white female cat, has just paid for the third recovering of the living room couch because her cat has repeatedly clawed up one corner of it. Unfortunately, this is the most visible part of the couch. This client does not believe in declawing cats and refuses to consider the operation. However, she is willing to pay for an office visit to obtain advice of a behavioral type.

Diagnosis. Scratching Furniture

General Evaluation. Tassle will be given an acceptable scratching post to replace the couch and the couch will be made less accessible.

SPECIFIC INSTRUCTIONS

1. Make the scratched couch unavailable by moving it aside and covering it temporarily with plastic.
2. Place a new scratching post at the spot where the couch corner was. Gently rub Tassle's feet on the post to deposit glandular secretions on the post. The post should be covered with material that has longitudinally oriented threads.
3. Once Tassle uses the post regularly, it can be moved to a less conspicuous location, inch by inch over several days.

the cat owner must have a piece of furniture reupholstered. As for the type of material with which to reupholster a chair or couch, the tightly woven knubby materials are the safest. Cats prefer materials with longitudinally oriented threads where a long stroke through the threads can help tear loose the old claw at the base. Naturally, if a cat does not have a satisfactory alternative for a scratching object, it may be forced to work over a less preferred material.

Many cat owners have noticed that scolding or hitting a cat for scratching furniture does not suppress scratching behavior. About the best that can be achieved is that the cat will run when the owner (punisher) approaches it. Punishment can be effective if one uses remote punishment, such as sneaking up on the cat and squirting it with a water sprayer, or hanging loaded mousetraps from strings above the inappropriate scratching area.

Cat owners usually have a couple of questions about cat scratching. One is the value of attaching catnip to the scratching post. If the catnip is effective, however, the cat will "trip out" around the post, which is hardly appropriate behavior.

Another question relates to the value of demonstrating to a cat where to scratch by rubbing its feet on the post. This is beneficial because this rubs off foot gland secretions that may help attract the kitten back to the post later.

Chapter 16

Territoriality and Roaming Problems in Cats

A cat's attachment to its home or territory becomes a matter of concern from two standpoints: (1) For some households a cat's roaming behavior may seem excessive, and the owners wonder what to do about it and if it is normal. (2) People sometimes find that their cat is more attached to their territory than they would like it to be. When they move to a new home that is not too far from the old one, the cat, preferring the old territory, may repeatedly leave the new home and return to its old territory. In this case, the cat's attachment to territory is stronger than its attachment to people. In other instances, one finds that as long as the owners are present the cat does not wander off or attempt to find the old territory. Understanding a cat's territorial behavior is often useful in handling some problems related to territorial attachment.

TERRITORY AND HOME RANGE

Usually a cat spends most of its time in a den or resting area, such as a particular room of the house, a corner of the garage, or the owner's bedroom. Clearly, the resting area is within a cat's territory, but it is difficult to determine how far a territory extends. An animal's territory is usually defined as the area that is actively defended. We are not usually able to define the actual territorial boundaries of cats unless we have observed and plotted the locations of fights.

Most outdoor cats, in addition to having a resting area and territory, have a home range. This area is covered by the cat in its normal day-to-day activities, but it is not necessarily defended. House cats may not have a home range beyond the territory. Cats from different households have home ranges consisting of yards and sometimes adjacent wooded areas. Home ranges of various cats often overlap and, therefore, are usually common pathways that are used by more than one cat. It appears that cats actually know their neighbors and are often acquainted with the movements or location of neighboring cats. Resident cats probably know about a new cat entering a neighborhood relatively soon. When several cats live in one household, each cat has its own lounging area, but they come together at feeding or other times.

Providing cats with shelter and food brings about a major change in the social behavior of cats, which would otherwise be asocial. Some type of social order is usually present to reduce major conflicts. In a study of the daytime behavior of eleven free-roaming cats on a farm, Laundre (1977) found that the cats fluctuated throughout the day between solitary and group living. Most of the social interactions and aggressive encounters occurred when the cats came together twice a day during feeding. At times other than feeding, most of them went off alone, some hunting, others just resting. Most cats had the same pattern day after day. The areas of some cats overlapped, and were used simultaneously by two cats, but no territorial disputes were seen during these times. With respect to social hierarchy, the two females were dominant, and the males initiated aggressive encounters so seldomly that it was not possible for them to be ranked. These cats preferred a solitary lifestyle most of the time but adapted to a social life style during feeding times.

Leyhausen (1965) studied free-roaming cats from several different households. These cats also used overlapping areas and a type of time-related territoriality. They also avoided each other on pathways and thus further reduced encounters.

Conflict Resolution

Although cats are considered to be asocial, they have the ability to adapt to confinement with other cats, and can be kept together in a relatively small colony pen where they appear rather indifferent to each other. When confined, conflicts are resolved by threats of one animal and withdrawal by the opponent. Conflict resolution is effective providing cats have some personal space. In close quarters the more social animals, such as dogs, would form a fairly rigid social hierarchy. However, most cats form dominance rank orders only to the point where there is a despot or leader (Figure 16–1). In some instances there may be a second ranking cat. The others remain subordinate to the leader, but indifferent to each other (Leyhausen, 1965; Baron, 1957; DeBoer, 1977).

Domestic cats in many neighborhoods often exist in a denser environment than if they were to space themselves out naturally. This higher density seems to present no difficulty for females, which do not wander much and are relatively nonaggressive toward each other. However, male cats that are not castrated have a greater tendency to roam and are much more aggressive toward each other and especially to strangers. Fighting often develops when there is a high population density, resulting in serious wounds. Prepubertal castration is usually effective in preventing roaming and fighting. However, prepubertal castration does not guarantee a home-loving, well-behaved male cat. In fact, the probability of a

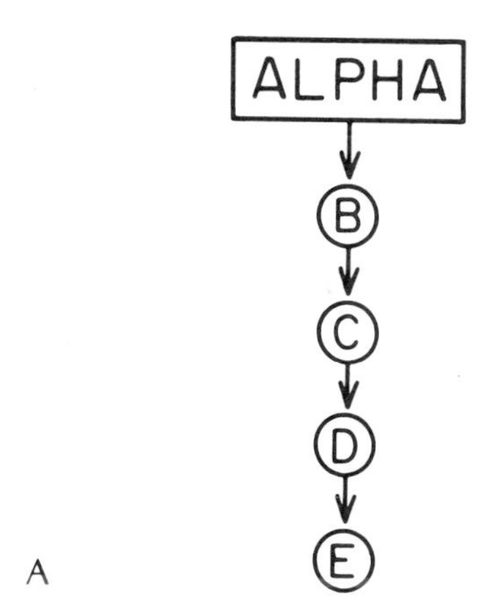

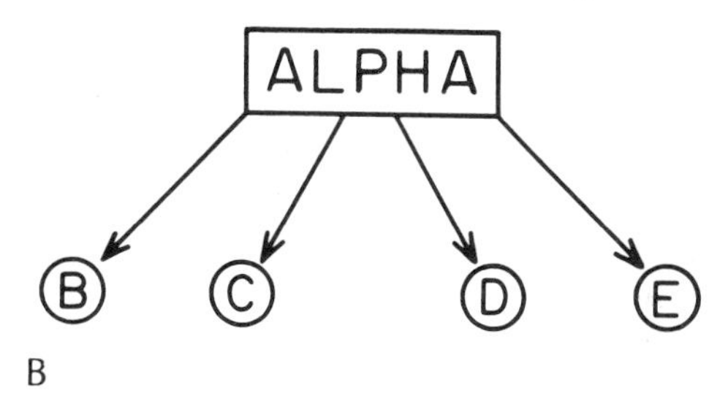

Fig. 16–1. The social dominance hierarchy system typical of dogs is depicted here as a linear hierarchy *(A)*, whereas that typical of cats is depicted as despotic *(B)*, with an alpha cat that dominates other cats who are basically indifferent to each other in dominance–subordination relationships.

prepubertal castrate fighting is about the same as that of a postpubertal castrate (Hart and Cooper, 1984).

Castration has been shown to be about 90% effective in reducing or eliminating roaming in adult tomcats (Hart and Barrett, 1973). If castration is employed to reduce roaming, one can expect about an equal chance of the change being rapid or gradual.

Traveling, Boarding, and Cat-sitting

A frequent question regarding territory is whether to take the cat along on vacation, or have a friend come to the house and feed it during the owner's absence. In most cases it is advisable for the cat to be left in its own home territory to be fed and cared for. The territory is usually more important than any social attachment the cat may have to a friend.

With the owners gone, the cats will at least be in a familiar environment. Because cats are basically loners, they do not show the same emotional disturbances as more social species, such as dogs, when isolated. If taken to a friend's home, the cat may become disturbed at the presence of a strange cat or dog. Also, a visiting cat may create some problems with the friend's cat, such as aggressive outbursts or spraying.

PROBLEMS RELATED TO TERRITORY AND ROAMING

Changing Homes

What can one do with the cat that keeps going back to its old territory? One approach is to confine the cat to the new home for a couple of weeks. In fact, temporary confinement should be practiced every time someone moves to a new home with a cat. A cat can adapt to its new outdoor surroundings by tying it outdoors with a harness for several hours at a time. While administration of a tranquilizer can be useful in calming a cat that is extremely upset in a new home, it is imperative that the tranquilized cat not be allowed outside where it might attempt to return to its old home and find itself in something less than full control and become lost or hit by an automobile.

When the cat does return to its old home, one can enlist the help of the occupants in discouraging the returning behavior by asking them to squirt water from a garden hose at the cat every time it shows up. Yelling and throwing things may even help. The worst thing is to reward the returning behavior by carrying the cat home or having the new occupants shelter and feed the cat until the owners return for it.

Roaming in Males

Tomcats are known for their roaming episodes. As mentioned above, castration is usually effective in preventing or eliminating roaming. If a castrated male is still roaming, the indicated therapy is administration of a long-acting progestin, such as medroxyprogesterone or megestrol acetate (see Chapter 23). As with a cat going back to its old home after the owners move, enlisting the help of neighbors in discouraging the cat from entering their yards may help induce an errant male to remain home.

Chapter 17

Predation in Cats

Cat owners and the public at large experience a variety of feelings about the predatory behavior of cats. People love cats for their ability to keep the population of mice and rats at low levels on farms and other places where rodents may be a serious problem. Cats that will capture and kill prey on the farm, even when they are not hungry, are especially valued. However, cats in a suburban setting that frequently bring home song birds, squirrels, or chipmunks, are often disliked for the same behavior. The cat that kills birds when it is not hungry, and does not eat the bird, may be hated.

A cat and its hunting tendencies are not easily divided. When predation is a problem, owners usually live with or manage the behavior by confining cats, or having them declawed. In selecting and raising an urban cat, however, there are things we can do to avoid this problem. By the same token, action can be taken to enhance the predatory behavior of cats living on the farm.

NATURAL PREDATORY BEHAVIOR

We owe much of what we understand about this kind of killing and hunting to Paul Leyhausen's observations on wild felids and domestic cats under seminatural conditions (Leyhausen, 1979). The natural prey of cats are rats, voles, and mice. These small rodents can be successfully hunted by a lone cat, whereas large prey can only be hunted with the cooperation of two or more predators. Domestic cats have been known to kill and eat a variety of prey, ranging from small grasshoppers to rabbits. Their diet also includes house flies, lizards, snakes, and squirrels, but the most common prey are mice and rats.

Prey animals often escape, especially if they are not taken by surprise. Leyhausen estimates that a cat makes about three attempts before it catches one mouse. Even so, a cat is more capable of catching small rodents that live in burrows than birds. With burrowing rodents, watching prey before pouncing on it is worthwhile. If a cat were to rush a mouse or rat as soon as it appeared at the entrance to its burrow, the rodent would simply run back into the burrow before the cat could nab it. The cat's strategy is to wait until the rodent strays away from its shelter. As a cat stalks, it also waits, slinking along a trail or ditch and then watching for the moment of opportunity. This waiting behavior can be counterproductive when applied to birds. The natural tendency of birds is to keep hopping while searching for food. When the bird hops along, the cat must also move and settle into its waiting mode. The bird may initiate this movement process several times and eventually fly away, usually without having seen the cat.

Studies based on the examination of fecal droppings, stomach contents, or prey delivered to the owner's home, indicate that only 4 to 18% of a cat's diet consists of small birds (Leyhausen, 1979).

On the other hand, the effect of feral cats on the rodent population can be very substantial. When the population of meadow voles surged in a northern California park, feral cats ate approximately 88% of the entire population (Pearson, 1964). Around homes where vines are growing on fences, or where yards border on woods that provide food for rodents, cats prey on the rodent population and keep it within reasonable limits.

Does hunger affect a cat's hunting tendencies? Cats are known to kill prey without immediately eating it, or they may kill more prey at one time than they can possibly consume. Through multiple killings and the tendency to kill without necessarily eating prey, cats maintain a sufficient food supply.

Most cat owners have seen cats play with their prey before eating it. Cats sometimes even delay delivering a killing bite as if to intentionally prolong their play with a half-live rodent. Prey is tossed into the air, batted around with the paws, rolled over, clasped and kicked in the stomach with the hind claws. One notion is that this play is a release of pent-up energies associated with different aspects of predatory behavior, a notion espoused by some behaviorists, including Leyhausen.

One of the fascinating aspects of early feline experience is the manner in which mother cats introduce kittens to predatory behavior. Yerkes and Bloomfield (1910) attracted attention when they claimed that cats needed no prior experience with prey to efficiently prey on mice. It was later shown that this was only true of some cats; others needed experience with prey as kittens to kill mice later. Caro's (1980a; 1980b) recent work indicates that cats exposed to mice as kittens gain in predatory efficiency, but with birds or fish there is little carry-over. Similarly, preying on birds does not improve a cat's skills with mice or fish.

If mothers are present when kittens are exposed to prey, the young cat's prey-catching behavior improves. Mothers first bring the dead prey to the kittens and eat it in front of them. Later the prey is presented to the kittens and the mother tries to elicit their interest in it without eating it herself. Finally, mothers provide their kittens with live prey and allow them to play with it, while not allowing the prey to escape. Gradually the mothers continue to diminish their role, phasing into the kittens capturing prey without assistance.

PREVENTION OF PROBLEMS RELATED TO HUNTING

Selecting Cats to Be Hunters

In rural areas and in homes where rodents are a problem, cats may be obtained for the purpose of rodent control. In these instances one wants to foster predatory behavior. The recommended procedures for encouraging predatory aggression must be taken before obtaining a kitten. Selecting kittens from litters where the mother is known for effective hunting, and then leaving the kittens with the mother long enough to learn to hunt, can assure that a kitten will probably become a good hunter. Many cats hunt even when they are not hungry, so it is wise for owners to feed their cats and thus encourage them to spend their time in the area where there is a rodent problem.

Selecting Cats to Be Nonhunters

People who live in urban and suburban areas and wish to select or acquire animals that are least likely to prey on song birds or small mammals should consider both the genetic and acquired aspects of predatory behavior. The genetic aspect may well be the most important factor. Selecting a kitten from a litter where the mother and sire are known to be nonhunters is the best way of assuring that a kitten will not become a hunter. If inquiries about siblings reveal that few or none of them are known to be hunters, then all the better. If the genetic approach is not possible, then a kitten should be removed from the litter early enough to prevent it from acquiring hunting skills under maternal tutelage.

Whatever the selection procedures, some cats do not become hunters even when the litter has been tutored by the mother to hunt. Other cats become hunters regardless of the fact that the mother did not hunt and the cats were not exposed to other cats who were hunters.

PREDATION PROBLEMS

If a pet cat manifests predatory tendencies that are a problem to the owner, one solution is to physically manage the problem. An outdoor cat that preys on neighborhood birds can be outfitted with a bell. This will not protect eggs or young birds in a nest, but it may increase the warning calls uttered by birds and reduce the vulnerability of mature birds.

If the predatory problem is an indoor one, involving caged birds or rodents, it may be possible with remote punishment and aversion conditioning to keep a cat away from specific areas where the potential prey is housed (see Chapter 22). A barrier of inverted mousetraps around a secure hamster or bird cage can guarantee that a cat will be punished each time it ventures near the cage.

Aversion conditioning can be used by obliquely squirting the cat near its nose with underarm deodorant. The aluminum chlorohydrate in most underarm deodorants is irritating to the mucous membranes of the nose and the cat may associate the perfume of the deodorant with the unpleasant, irritating aluminum chlorohydrate. Two or three exposures, a day apart, may be needed. Then spray the area around the cage at risk with the same deodorant. Have the client use an underarm deodorant different from his or her own to avoid producing an aversion to the owner. Then the cat may be allowed access to the cage to see if the aversion conditioning works. A simultaneous combination of aversion conditioning and remote punishment is the most powerful treatment. Whereas a cat might successfully adapt to a single technique, the combination avoids adaptation. Both remote punishment and aversion conditioning were included in the client's instructions for the sample case of Sarah and the canary (Case 17–1).

CASE 17–1

History. Sarah, a female Burmese, persistently tries to attack the canary that is kept in a cage on a bookshelf. The owner wants some advice on how to make the cat leave the canary alone.

Diagnosis. Predatory Behavior

General Evaluation. Remote punishment will be used to produce an aversion to the area around the canary's cage.

SPECIFIC INSTRUCTIONS

1. Make sure that the canary is as safely housed as possible.
2. Booby trap the area around the canary cage with inverted mousetraps so that Sarah will be punished each time she goes near the canary.
3. If the above does not work, obliquely squirt Sarah a couple of times near the nose with underarm deodorant that is not yours, and spray the access route to the canary (rug, shelf, etc.) with the same deodorant. This may give Sarah an aversion to the canary.

Chapter 18

Feeding and Related Problems in Cats

Although hunting is the mainstay of a cat's existence in the wild, on the domestic scene cats are provisioned with commercial food, and most of them maintain their weight and health at optimal levels. However, some cats are very finicky or selective in the food they eat, and others eat so much that they become obese. Some experimental work bears on these aspects of feline feeding behavior and aids in the understanding and treatment of clinical problems.

NATURAL FEEDING BEHAVIOR

When considering the evolutionary origin of feeding behavior in domestic cats, we often think of the house cat as related to the felids that prey on large grazing ungulates. This gives us the concept of cats experiencing a "feast or famine" type of existence, in which one kill may provide them with enough food to last for several days. The immediate ancestor of the domestic cat, however, is the small North African cat which preys on small rodents much as do feral cats on farms. For the mainstay of its diet, the cat's wild ancestor preyed on a rodent the size of a mouse. The caloric value of the usual ration of canned cat food is equivalent to that of one mouse. Most cat owners feed their cats just once or twice a day. However, cats will nibble throughout the day if given the opportunity. In a study by Mugford (1977) in which groups of adult cats were given free access to a complete cat food, the cats distributed about 13 meals evenly throughout a 24-hour period. Whether the food was canned, semimoist or dry, it had little effect on the frequency and spacing of meals.

Cats that feed exclusively on a single food, even though nutritionally balanced, seem to develop a transient depression of interest in that particular food and tend to favor an unusual diet. Providing animals with increased feeding opportunities and a greater variety in their diet can stimulate an increase in caloric intake (Mugford, 1977). The notion that cats and other animals should eat any nutritious food placed before them, if they are hungry enough, is popular. In some instances this may be an appropriate approach, however, in the hospital or cattery setting this is not reasonable. A rigid adherence to the principle that an animal will eat if hungry enough could cause malnutrition in sick or convalescing animals.

PROBLEMS WITH FEEDING BEHAVIOR

Cats can manifest problems of eating too little food. When they eat too little of everything, it is termed anorexia. If they simply avoid particular foods, they are said to have aversions to those foods. They can also eat inappropriate material, a condition referred to as pica, and they may also eat too much, becoming obese.

Anorexia

Not eating is a common sign of many illnesses. Some cats cannot tolerate too many days without eating, so persuading a cat to eat becomes a particular concern. It could be argued that this behavior, when coupled with a fever response, helps cats recover from some diseases. The anorexia allows a cat to stay inactive and maintain an elevated temperature which is inhibitory to the growth of microorganisms. The anosmia that accompanies some types of upper respiratory diseases is one of the causes of anorexia. Force feeding a cat, so that taste receptors are activated, is one way of getting such cats interested in eating again.

Interest in food may be stimulated by administration of a steroid. The progestins that are regularly used in behavioral therapy often markedly stimulate appetite as a side effect, and might be given for their appetite-stimulating effects alone (see Chapter 23).

Lack of interest in some foods may relate to where cats are fed. Some cat owners, who wish to confine the mess on the kitchen floor, place the food and water dishes next to the litter box. Certainly it is natural to expect an animal to be repelled by having to eat next to its own eliminative area.

Food Aversions

Cats may refuse to eat because they are not interested in food, or they may have an aversion to particular foods. Sometimes a lack of appetite may reflect an acquired food aversion stemming from the occurrence of a gastrointestinal illness after eating a particular food. The most likely example in domestic cats would be an aversion to a food which produces an allergic reaction.

In research laboratories the aversions most frequently studied are those in which a particular food has been paired with a drug that produces a transient gastrointestinal illness. If the treatment causes post-ingestional nausea or sickness, the taste or smell of the food associated with the treatment becomes aversive after one or more pairings. This type of food aversion is common among people. Almost everyone can remember a food that was previously favored, but later became aversive after the food was associated with a gastrointestinal illness or nausea.

Laboratory studies have shown that cats can acquire specific food aversions. Mugford produced an aversion in cats to a highly palatable meat-based canned diet by treating the cats with lithium chloride immediately after they were fed, and the aversion persisted for at least 40 days.

The adaptive role of this type of aversion learning is in protecting wild felids from repeated ingestion of prey, or organs of prey, that might produce gastric distress or nausea due to endotoxins. With one exposure the cat will avoid the source of the toxins. An aversion is also one way the body can protect itself from repeated bouts with food allergies. A number of drugs including some tranquilizers can also produce food aversions if they are paired with food. Thus, the sudden aversion for a food by a cat might be related to repeated administration of the drug in a particular food.

Inappropriate Appetites

Sometimes cats exhibit excesses in their feeding habits. This can include feeding on an inappropriate material, such as wool or a houseplant.

Wool Chewing

Just the opposite of an aversion is an attraction of an animal to an unnatural nonfood material. Wool chewing in cats is an interesting aberration of ingestive behavior. This behavioral trait, generally restricted to the Siamese breed, begins at about the time of puberty, when cats may suck, chew and/or ingest chunks of wool from stockings, sweaters or caps. Although wool is usually preferred, many cats generalize to other types of cloth and even synthetic fibers. Some cats threaten to eat their owners out of the house and become such a problem that they cannot be kept as indoor pets. Most, however, seem to give up wool chewing within a year or two. This behavior, like other forms of pica, has not been explained and is not understood. No treatment has been reported as successful, although aversion conditioning with underarm deodorant, as outlined in Chapter 22, may prove effective.

Plant Eaters

Eating plant material, especially grass, is a natural behavior for cats, but if a cat munches on a favored houseplant, it can become a serious problem to the owner. While the function of plant eating is not understood, the behavior can sometimes be prevented by providing a small garden of grass for the cat. Once a cat has the habit of eating houseplants, remote punishment each time the cat goes after the houseplant is most effective (see Chapter 22). This may require limiting the cat's access to plants to times when the owner is nearby with a water pistol. Inverted mousetraps require somewhat less vigilance by the owner.

Aversion conditioning with underarm deodorant can sometimes make houseplants less attractive to cats (see Chapter 22). The case of Fonz (Case 18–1) illustrates instructions to a client when a cat is eating houseplants.

Obesity

The critical issue in weight control is whether caloric intake balances energy output. Some cat owners frequently feed their pets very palatable foods in large quantities, even though the cat does not have to expend calories to hunt or search for food. In treating overweight animals, the critical issue is the restriction of food intake. This of course requires that the owner manifest self discipline and restraint.

Cat owners often look for physical causes of their cats' obesity. Questions raised include the possibility of a hormonal imbalance, or the influence of spaying or castration. Aside from the obvious example of thyroid insufficiency, little is known about the effects of hormones on body weight in cats.

In laboratory rats an ovariectomy leads to a pronounced increase in body weight, along with an increase in food intake. Estrogen administration reverses this effect. It is questionable whether this finding in rats regarding estrogen relates to cats since a female rat manifests an estrus

CASE 18–1

History. Fonz, a domestic shorthair castrated male cat, eats leaves from several of the houseplants belonging to the owner. Most of the eating is done when the owner is gone, but sometimes the cat eats plants when the owner is in the same room but not looking.

Diagnosis. Eating Houseplants

General Evaluation. Fonz will be given remote punishment each time he approaches a houseplant.

SPECIFIC INSTRUCTIONS

1. Allow Fonz access to a small grass garden.
2. Booby trap all houseplants by surrounding the area around the bases with inverted mousetraps. It is crucial that Fonz receive punishment whenever he approaches a plant.
3. If some plants are impossible to booby trap, or if there are too many to booby trap, relocate some plants in a room where Fonz is not allowed.
4. When you are around and see Fonz approach a plant, also secretly ambush him with a water pistol.
5. If the booby trapping and ambushing do not work, attempt aversion conditioning by obliquely squirting Fonz a couple of times near his nose with underarm deodorant that is different from yours and spray the houseplants with the same deodorant.

cycle every 4 to 5 days throughout the year, and is more subjected to estrogenic influences than a cat cycling seasonally. Even so, the work on rats and dogs (see Chapter 23) raises the possibility that a slight increase in body weight may be attributed to a gonadectomy.

Progestins are also known to influence body weight gain. In clinical observations, cats treated with a long-acting synthetic progestin such as medroxyprogesterone or megestrol acetate often exhibit a pronounced increase in appetite and food intake.

There is growing evidence of a physiological basis for obesity in some individual animals and people. Whether through inheritance or by being overfed as infants, some animals have a larger number of fat cells than others. The number of fat cells remains constant in the adult, and this may account for much of the difference between obese and normal individuals (Nisbett, 1972). Perhaps some cat breeders, in their concern to keep young kittens as healthy as possible, excessively supplement the mother's milk. This would cause hyperplasia of fat cells in the kittens and facilitate permanent obesity. Obese individuals may simply have a higher biological adipose set-point. Forcing such an animal to reduce may be analogous to taking an animal of normal weight and starving it.

Undoubtedly, the main cause of obesity in cats is simply that the owners offer their cats too much highly palatable food, and give them too many calories for the minimal amount of exercise they get. Without having to hunt or forage, cats expend less energy each day than their wild ancestors. The physiological regulatory system is not refined enough to deal with modern urban living in cats anymore than it is with modern urban living in people.

Chapter 19

Sexual Behavior Problems in Cats

More than any other area, cat owners look to veterinarians and other professionals for behavioral advice regarding reproductive behavior. Sexual behavior in particular is closely aligned with medicine and physiology. It is an area that many people feel embarrassed about, but expect professionals to be competent in discussing. It is important not only to be able to deal with problems in this area, but to be able to clearly differentiate normal from abnormal behavior. This chapter deals first with normal feline sexual behavior and then with some of the problems.

Cats are, to a large extent, nocturnal, which means that much of their general activity takes place at night, including interactions with the opposite sex. Most cat owners, therefore, are less familiar with the sexual behavior of cats than of dogs, since dogs are much easier to observe. Many veterinarians have been called at late hours by anxious feline owners wondering why their female cat is acting "crazy."

Objectionable sexual behavior, such as mounting people or other animals, is a concern, along with breeding problems.

NORMAL SEXUAL BEHAVIOR

When a female cat is in estrus her behavior is very distinct, including heightened activity and nervousness. Her mating call apparently attracts male cats, and sex attractants in her urine stimulate visiting male cats to stay around. Unfamiliar male cats may appear on the doorstep of the owner of an amorous female and emit courtship cries, unequivocally communicating their interest to the female. In the presence of a male cat, or even if she can only hear or smell him, the female is likely to assume the receptive posture, consisting of elevation of the pelvic region, deviation of the tail to one side or the other, and treading or stepping of the back legs (Fig. 19–1*B*). As the male cat investigates and starts to mount the female these responses usually become more intense.

Female cats differ from females of other domestic species in that this receptive behavior is sometimes displayed to the pet owner. The pelvic

Fig. 19–1. Elements of mating behavior in cats. Upon genital investigation males may display flehmen behavior *(A)*, which presumably involves the accessory olfactory system (vomeronasal organ) in detection of sex pheromones. The receptive posture of female cats, consisting of pelvic elevation, treading of the back legs and tail deviation, are displayed to tomcats as they approach a female *(B)* and make their initial mount, which includes a neck grip *(C)*.

Fig. 19–1. *Continued* As the male begins pelvic thrusting, he slides posteriorly while maintaining his neck grip, until genital contact is made and he engages in intromission, which lasts only a second or two *(D)*. At this time the female typically becomes highly excited, emits a copulatory cry, and will frequently turn and swat at the male *(E)*. The female next grooms her genital area *(F)*

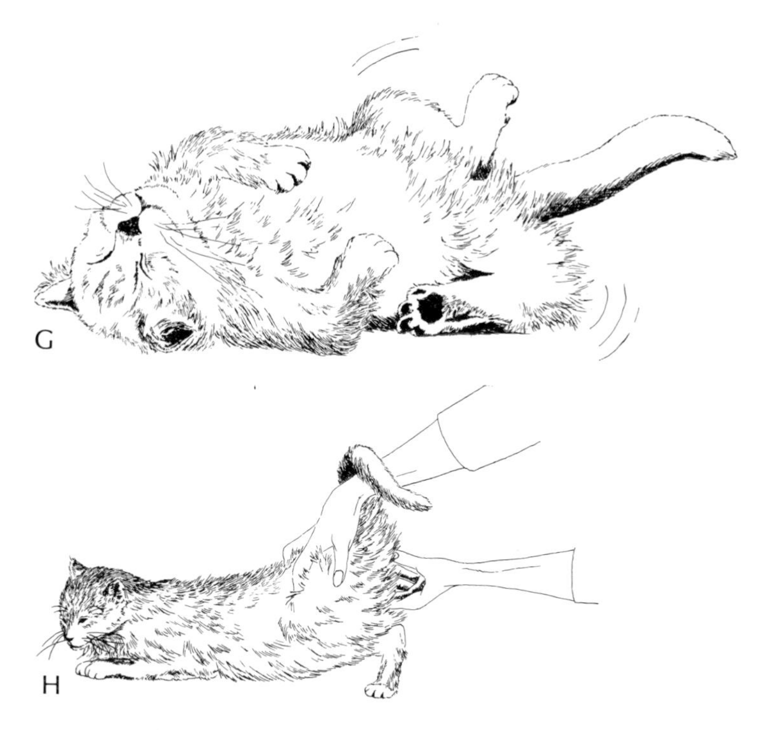

Fig. 19–1. ***Continued*** and rolls and rubs on the floor in what is termed the mating "after reaction" *(G)*. As a test for estrus in female cats, one may evoke the same posture displayed towards male cats by perineal stimulation *(H)*. This behavior is displayed intensely only if the female is in estrus.

elevation, tail deviation, and treading can often be induced by human handlers stroking the back of the female cat and touching the perineal region. The response may be intensified by grasping the skin over the back of the neck while stroking the perineal region with the other hand (Fig. 19–1*H*).

If a male is comfortable with the surroundings, he will then approach the female, usually engaging in nose-to-nose as well as genital investigation. This often evokes a flehmen response in the male (Fig. 19–1*A*). He then proceeds to take a neck grip on the female with his teeth (Fig. 19–1*C*). The male usually engages in treading or stepping of the back legs after he mounts. His initial mount is usually fairly high on the female's back, but he generally slides backward over the female as he continues leg stepping until he is aligned for intromission (Fig. 19–1*D*). Leg treading by the female also aids in bringing about genital contact. The male then begins pelvic thrusting, and copulatory intromission is followed by a deep pelvic thrust. The male remains motionless for a few seconds after intromission, and during this time a degree of excitement seems to build up within the female as her eyes dilate. Soon after ejaculation she begins to pull away from the male and almost simultaneously emits a loud cry. She turns to hit at the male as he springs back (Fig. 19–1*E*), and she quickly begins licking her genitalia (Fig. 19–1*F*) and going into the after-reaction which consists of rolling and rubbing on the floor (Fig. 19–1*G*).

Since the female cat is a reflex ovulator, the duration and number of estrous periods during the breeding season is affected by the frequency of mating. When she is mated, estrus lasts for only four to six days, but if she is not mated, estrus may last for as long as ten days, and subsequent estrous periods may occur at intervals of two or three weeks. Ovulation can be induced by probing the vagina of a female cat with a smooth blunt instrument such as a glass rod. Several insertions of about 10 seconds duration and 5 minutes apart for two successive days will usually induce ovulation (Diakow, 1971). Often females will display the copulatory after-reactions to these insertions. Hence, in this way repeated or prolonged estrous periods can be avoided if ovariohysterectomy or breeding is not desired.

THERAPEUTIC APPROACHES TO PROBLEM SEXUAL BEHAVIOR

Problems with Males

There are two categories of problems: objectionable behavior and lack of sexual ability or interest in breeding males.

Objectionable Mounting of Other Cats

On occasion castrated males may persistently mount other males in the household or female cats, even those not in estrus. One approach is to reduce the motivation of the mounter to mount. The indicated treatment for this is a long-acting progestin such as medroxyprogesterone acetate or megestrol acetate (see Chapter 23).

The second approach is based on the principle that a cat would not mount other cats if they did not tolerate it. Remote punishment with a water sprayer (see Chapter 22), might prove effective if the cat being mounted is sprayed each time it allows mounting, and if the spraying is done so that the cats do not know who is delivering the punishment. When no one is around to deliver remote punishment, the cats should not be allowed together until the problem is solved.

Inadequate Sexual Interest of Breeding Males

Male cats may lack interest in copulating with a receptive female for a variety of reasons. One is discomfort with an unfamiliar breeding environment. For some male cats a strange environment seems inconsequential, but others require a month or two after being brought into a new environment before they will readily mate. Even a male cat that readily mates when a female is brought to him may not initiate mating in a new environment.

Another type of problem is a male cat that seems incapable of obtaining complete intromission, even though he persistently mounts and thrusts. Lack of experience in executing the right movements may result in this problem. The only therapeutic measure is to provide the male with more experience.

Hair Rings

A hair ring sometimes develops around the glans penis, preventing intromission and, hence, successful mating (Hart and Peterson, 1971). One clue that this problem exists is a male that continually thrusts but does not intromit. The glans penis of the male cat is covered with epithelial papillae which project backward, and the papillae apparently act to collect hairs (Fig. 19–2). The hairs may come from the preputial sheath or, perhaps with frequent mating, from the fur of the female when the erect penis is rubbed over the back and perineum of the female. The hair ring is sometimes removed by the males themselves. One can also remove the ring by gently sliding it over the penis. Animals are able to mate immediately after removal of the hair ring, and the rings do not appear to be painful.

Testosterone Levels

A male's lack of interest in mating could theoretically be due to abnormally low levels of testosterone. One way to evaluate male cats in this regard would be to analyze blood testosterone levels and compare these to a normal range. However, it would be necessary to submit several blood samples taken throughout the day, since testosterone blood concentrations fluctuate. Evidence at present indicates that less than half of the usual level of testosterone can maintain normal copulatory activity in male cats. Thus, testosterone levels would have to be extremely depressed to account for a male's lack of sexual interest in females (Hart, 1979e).

One area where testosterone levels may be important is in regulating sexual behavior during seasonal fluctuations. Differing from dogs and other domestic species, male cats exhibit evidence of seasonality in sexual behavior. For males, the lowest period of sexual activity is in the fall. One would not expect this to be an important factor in breeding, since females usually are not in estrus at this time. However, a seasonal fluctuation should not be automatically disregarded in a consideration of an otherwise unexplained decline in the sexual prowess of a breeding male.

Training a Breeding Male

Once cat owners have decided they have a male worthy of carrying on the family name, they may find to their disappointment that he lacks the sexual prowess equal to the task. A few precautions may save the

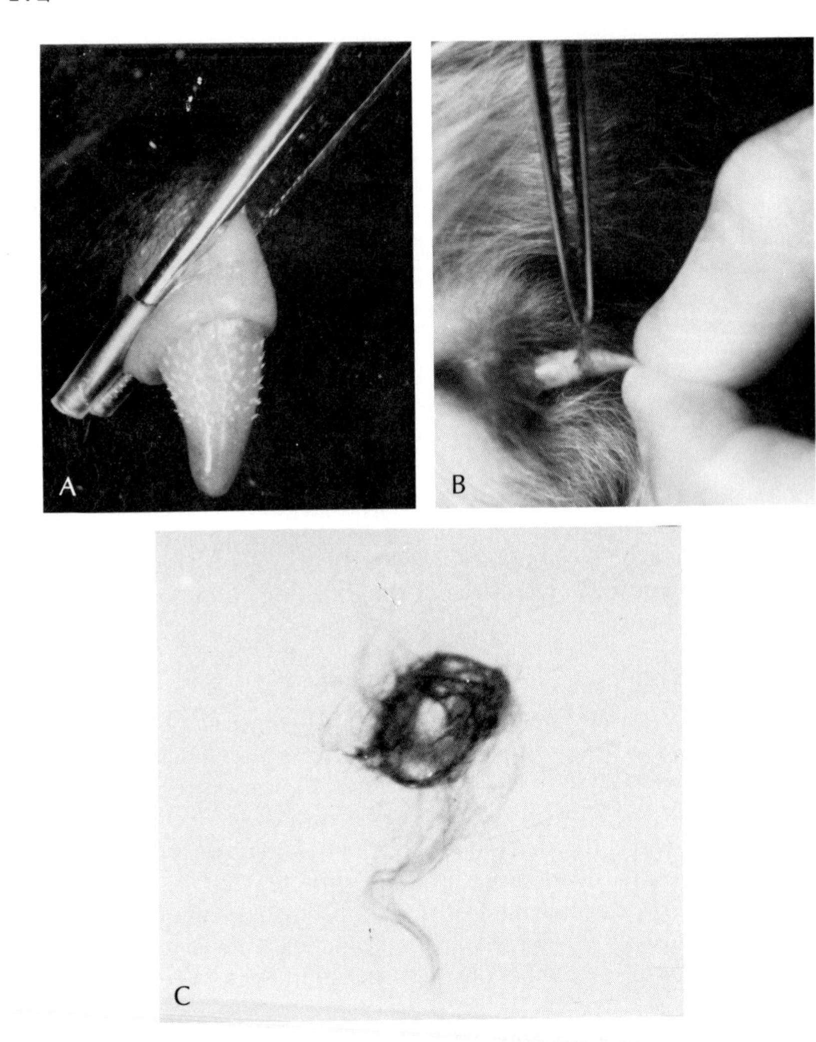

Fig. 19–2. Epithelial papillae or spines *(A)* that cover the glans penis collect hairs, and these can cause a hair ring to form around the penis *(B)*, which prevents intromission. The removed hair ring appears in C. The impairment in intromission is difficult to distinguish from a behavioral impairment. *(B,* from Hart and Peterson, 1971; reprinted with permission, American Association for Laboratory Animal Science.)

owners some dismay (Hart and Voith, 1977). Providing frequent copulatory experiences with very receptive females can condition many male cats' breeding activities. These should be comfortable occasions in a standard environment, with a routine sequence of procedures. If a specific area is reserved for breeding, the male cat will come to anticipate copulation when he is placed in that area. For each mating occasion, the male should be given time to acclimate to the area, and after a few minutes a receptive female placed with him. The male is allowed to mate with the female several times in succession if he is so motivated. During the initial sessions, the male may wait 30 minutes to several hours before copulating. However, usually after a few weeks of frequent presentations of a receptive female, he will copulate within 15 minutes or less.

Sexually receptive females should be used when training a breeding male. Since female cats show a wide range of intensity in receptive behavior, an owner may need to test more than one female to find the most appropriate one to bring into behavioral estrus for training purposes. Two doses of .25 mg estradiol cypionate (Upjohn) administered subcutaneously two days apart will usually initiate estrus in a queen 24 to 48 hours after the last injection; and she will remain in heat for 14 to 28 days. Spayed females are best to use for training sessions to reduce the risk of uterine abnormalities. It is recommended that a female not be continuously used for several months because of the possibility of depression of bone marrow hemopoietic function.

Problems with Females

In order to breed females, it is crucial to carefully pinpoint the time of estrus. If the owner carefully observes the queen's behavior or listens to tomcats in the neighborhood, he will usually find it easy to know when the time has arrived.

Problems with females fall into two categories: difficulties with estrus detection in which one feels the female is overdue for coming into estrus but shows no behavioral signs, and rejection of a male's sexual advances even though the female shows other signs of sexual receptivity.

Estrus Detection

Females which the owners desire to breed do not usually live with sexually active males, and the owners are thus faced with the problem of determining when the queen is in estrus. Unlike the dog, the female cat does not have noticeable vulva swelling or vaginal bleeding as an indication of impending estrus.

Of course the sexual attractants emitted by females can attract wandering tomcats to the house, providing a clear signal. Frequently this takes the form of sex vocalizations and rubbing and rolling on the floor. With many female cats the behavioral signs associated with sexual receptivity are an extremely useful indication of estrus. These signs may be evident when the female hears or smells a tomcat. Petting on the back will often cause the female to crouch and elevate the pelvis. Although rubbing the perineal region while holding on to the skin over the dorsum of the neck (as the male cat would do) frequently evokes treading of the back legs and deviation of the tail to one side, such responses cannot be evoked from all cats in full estrus. Therefore, it would be a mistake to rely on this behavioral response as the sole indicator of estrus. One may have to place the female in the vicinity of a sexually active male to observe behavioral signs of estrus.

Rejection of the Male

A complaint regarding breeding females is that a queen that appears to be in full estrus will not accept the male. Restraining the reluctant female should be tried in these cases. Even though some male cats are reluctant to mate with a restrained female, an experienced and regularly active stud will often breed a female restrained by the breeder. If the restraining does not work, one might resort to leaving a male and female together for several hours with the hope that mating will occur. The drawback with just putting animals together is that one can not be sure if mating has occurred. If the female remains in estrus for a prolonged period after this type of mating, she presumably has not ovulated.

Some females may reject one particular male but accept others. If it is thought that the female is responding differentially to different males, then it would be advisable to try the female with a different stud male.

The area of breeding is fraught with individual idiosyncrasies and special sensitivities. From the standpoint of our management of this aspect of a cat's life, persistence and patience are invaluable.

Chapter 20

Maternal Behavior Problems in Cats

Of all domestic animals the cat is the most capable of going through parturition and the raising of its young without human care or intervention. In the classic instance, a mother cat reveals a litter of perfectly healthy kittens to its owners some 6 weeks after they have been born. This romanticized view of feline motherhood has given way to the more formalized cattery operation, or the family situation where a mother cat may give birth to her litter of kittens in the midst of an overly concerned family audience.

Problems with maternal behavior manifest themselves as lack of proper attention to the kittens, resulting in suffocation, undernutrition, hypothermia, or killing of kittens which can be accompanied by cannibalism. As with other areas of problem behavior, it is best to begin this section with a discussion first of normal behavior.

NORMAL MATERNAL BEHAVIOR

Maternal behavior includes several aspects of central importance in the life of a mother cat. For an infant kitten, not only is maternal behavior necessary for survival, but the experiences the kitten receives as a result of maternal attention and interaction with littermates have a critical influence on its behavior and health as it grows into adulthood.

As with dogs and other domestic mammals, we have taken over some aspects of maternal care by providing food, water, and shelter to kittens when maternal care has not been adequate, leading to the survival of cats whose mothers were lacking in the genetic programming of maternal behavior. Our intervention helps perpetuate the genes for poor mothering. The variety of maternal attitudes among cats is especially evident to managers of catteries. Some cats never seem interested in their own kittens, even after several litters. Others are wildly interested in all kittens they come in contact with.

Behavior Before and During Parturition

As the time of parturition nears, pregnant females become less active, and licking of their abdominal and genital areas increases. Shortly before parturition some females may seek a dark, dry, and relatively undisturbed area where they will deliver their young. Domestic cats that are strongly attached to their owners may not choose a location isolated from them. In fact, some cats become quite emotional during parturition if the owners are not nearby. Cats may become irritable or actually aggressive as the time of parturition nears.

The four stages of labor are: contraction, delivery of the fetus, delivery of the placenta, and the interval between deliveries. The interval between deliveries includes initial maternal care and leads into the immediate postparturient care of the young.

Uterine contractions begin in the first stage and there is a good deal of straining. Cats usually lie down during this stage of labor, although they may frequently sit up to change positions. Contractions of the uterine and abdominal muscles become more intense in the second stage, and the fetus moves rather rapidly through the birth canal. The female often breaks the fetal membranes with her teeth when the head or buttocks of the fetus appear at the vulva. She may actually pull the fetus through the birth canal by tugging on the membranes. Her typical posture is lateral recumbency with her head bent to the hind quarters through her back legs. After the newborn has passed through the birth canal, the mother rapidly consumes the fetal membranes and begins licking the newborn vigorously, which usually causes the first respiratory movements.

During placental delivery in the third stage, the mother continues to lick and groom the newborn. The placenta is usually eaten by the mother as it is passed. While the mother eats the placenta, she generally bites off the umbilical cord. The pulling and stretching involved in eating the placenta and umbilical cord seem to cause vessel constriction of the cord. Occasionally, movement of the newborn and the mother causes breakage of the cord. At other times the umbilical cord does not get broken within a short period after birth, and it is necessary for a person to intervene.

Between deliveries the mother not only continues to lick and groom the newborn animals, as well as her own genital region, but also cleans the bedding that has been soiled with amniotic fluids. The mother, according to actual observations, may lick her own body more than that of the newborn. Licking behavior seems to be less a response to the newborn than a response to the presence of fluids (Schneirla, Rosenblatt, and Tobach, 1963).

There is a wide (normal) range of variation in the duration of stages in kitten birth. The stage of contraction ranges from 12 seconds to 1½ hours, and the stage of delivery from 30 minutes to 1 hour (Schneirla, *et al.*, 1963). The birth sequence does not seem related to the duration of individual kitten births.

Only minor differences have been reported in parturient behavior between primiparous and multiparous females. Experienced mothers are less disturbed by physiological changes during birth and appear to respond more readily to the neonates in licking, grooming, and retrieving.

Immediately after the birth of the last kitten, a female lies almost continuously with her young for 12 hours or more. Newborn kittens usually begin to nurse within an hour or two after delivery of the last fetus. The mother cat remains almost constantly with her litter for about the first 2 days and only leaves the nest for very short periods to move about and feed. She takes breaks away from the nest more frequently after these first 2 days. Later, the amount of time the mother spends nursing the young relates to the size of the litter. A mother may spend 70% of her time nursing if she has a litter of several kittens, but with only one or two kittens, it may be considerably less.

Nursing

For the first 3 weeks, the mother initiates essentially all nursing sessions. While hovering over the litter or lying near them, she licks the kittens and arouses them. She typically lies with her body arched around the litter with her teats exposed. With time the newborn become very adept at finding teats and responding to the mother's solicitous behavior.

Many kittens are able to take specific nipple positions with some regularity after just 2 or 3 days of life. Others do not seem to prefer a particular nipple location. From about the third week of life, the eyes and ears of kittens function well, and they are able to leave the nest and to recognize and interact with the mother outside the nest. The young now initiate most nursing episodes. These episodes take place inside the nest or outside. The mother generally cooperates by immediately lying down and making her nipples available.

From about the fifth week of life until weaning, the kittens initiate all nursing. As time progresses the mother begins evading the nursing attempts of the young. For example, she may lie with her mammary region against the floor or climb up on objects so that the young cannot reach her. Weaning occurs near the end of this phase by the mother becoming less available to the young. At the same time, the young are becoming more capable of taking adult food. Mothers of wild feline species typically provision the young with rodents near the end of this phase.

The mother cat continues to lick and groom the newborn through the first 3 weeks. The emphasis on grooming the anogenital region evokes elimination, and the urine and fecal material are consumed by the mother, thereby keeping the nest clean. As the young begin leaving the nest area, the anogenital licking subsides, and the young deposit feces and urine away from the nest in another part of the room or another side of the nest box. The mother continues to keep the nest area clean.

Retrieving Behavior

The classical sign of interaction between a mother and her young is the sight of a mother carrying a kitten by the nape of the neck. The

Fig. 20–1. This posture, and the passive immobility induced by grasping a cat by skin over the back of the neck, is the same posture assumed by kittens being transported by a mother cat. The immobility is sufficient for giving injections or taking rectal temperatures in unruly cats. (From Hart, 1975a; reprinted with permission, Veterinary Practice Publishing Co.)

immobility posture that the young assume when being carried by the nape of the neck may also be induced in adult cats. In fact, minor procedures such as giving subcutaneous injections and taking rectal temperatures may be performed on cats restrained in this fashion (Fig. 20–1).

The tendency to retrieve peaks at about one week after parturition (Schneirla, *et al.*, 1963). Mother cats retrieve when they hear vocalizations of the young, particularly when the sound reaches a high intensity. This is why kittens that are marooned several feet from the nest and emit stress vocalizations are so readily retrieved.

The tendency for a mother cat to shift her litter from one spot to another in response to environmental disturbances is well-known. This tendency to move a litter is strongest between 25 and 35 days after birth (Schneirla, *et al.*, 1963).

Adoption of Strange Kittens

The ease with which some mother cats will lavish care upon strange kittens is quite surprising. When two or more females in a house or cattery have given birth to kittens around the same time, one is likely to find the mothers stealing kittens back and forth or harassing each other for the other's kittens. This behavior may result in a type of communal nursery with all the kittens piled together, while mothers take turns caring for the kittens, or all jointly lying with the kittens. This behavior would not be demonstrated by the cats' wild ancestors, since they lived in a solitary fashion. As counterproductive as raising offspring from another mother would be, natural selection did not act to produce a rejection of other kittens, since strange kittens were never present.

PROBLEMS WITH MATERNAL BEHAVIOR

Maternal reactions that create problems fall at the emotional extremes, ranging from attacking and often eating newborn kittens, to ignoring the kittens and allowing them to die.

This is the place to emphasize that the tendency of cat owners to intervene and help mothers with inadequate maternal behavior should be balanced against the knowledge that aiding the survival of young from mothers that provide inadequate care removes the selection pressures against poor mothering in the feline population. The result is a high degree of variability in mothering behavior as the offspring of poor mothers live to reproduce at almost the same rate as the offspring of exemplary mothers. The rather incomprehensible occurrences of maternal neglect or cannibalism can be extremely upsetting for clients. Your role is to cast all aspects of maternal behavior in a biological perspective. Cats cannot be blamed for being poor mothers any more than they can be blamed for being poor hunters or for not being outgoing and friendly. A main concept to get across is that maternal behavior is at least as much a reflection of genetic predisposition as it is learning or experience. While some cat mothers improve with experience, cat owners should be discouraged from feeling that they are likely to be able to teach their cats to be good mothers.

Maternal Neglect

Some observations on disturbances in maternal behavior have been made at a cat colony in Scotland by Dr. M.F. Stewart of Glasgow Veterinary School. In one of Dr. Stewart's surveys, about 8% of the kittens born in an apparently healthy state died, mostly from causes attributed to inadequate or inappropriate maternal care. Kittens may die of hypothermia because the mother does not remove the fetal membranes and dry the kittens, or she does not keep them close to her. This may occur because a kitten is sickly and unresponsive. Also, a mother cat may lie upon kittens and smother them. Failure to remove fetal membranes may reflect a lack of interest of the mother, or that she was distracted while cleaning the newborn. In large litters tangled umbilical cords may occur as the kittens arrive quickly and the mother does not clean off the fetal membranes. Hypothermia occurs if kittens are allowed to remain outside the nest and the mother fails to retrieve them. Once in a while a kitten gets dropped outside the nest because it held onto a nipple when the mother left the nest. Hypothermia also occurs when the mother does not stay with the litter. Stewart points out that some mothers may attend to a kitten in the nest and not notice that others are lying alone on the cold floor. Stranded kittens or kittens that a mother has abandoned can be gently warmed and presented to the mother again. Stewart has found that sometimes such kittens may be accepted by the mother and recover completely. Others are rejected by the mother after being repeatedly presented to her.

Stewart has some recommendations for the management of catteries. The mother should be in an individual parturition box at least three days before she is due so she can adapt to new surroundings. The queening box should be stable so it will not be knocked over by the mother and the sides high enough to prevent kittens from falling out. At the time of parturition mothers should be checked to see that they are cleaning the kittens and that the kittens are suckling soon after birth. Mothers that have presented problems in the past should be watched closely.

Cannibalism

A mother cat that is found to be cannibalistic usually appears alert and attentive to the kittens in every other respect. Circumstances that have been reported to be related to the occurrence of cannibalism are a litter larger than usual, second pregnancy of the season, and the presence of kittens who are ill. Previous experience in being a mother appears not to be related to cannibalism.

Perhaps the most accurate explanation of cannibalism is that the ability to kill newborn and eat them is a genetically programmed behavior that would appear to have an adaptive role in nature. If a kitten is sick and likely to die from disease or a congenital defect, the mother by killing and eating it keeps the nest from becoming soiled by a sick or dead kitten. By reacting to what could be an early sign of disease, the mother may protect the other kittens by promptly removing the slightly sick kitten before it is seriously ill. Rather than just depositing the dead kitten outside where it could attract scavengers, the mother, by consuming it, removes the evidence, and at the same time gains some additional nutrition, which means she will be gone from the nest a little less to replenish her own nutritional resources. The fact that a mother's cannibalism must be triggered by the very first sign of illness in a kitten, such as inactivity or hyperthermia, means that a disturbance of the environment could affect kittens and be interpreted by the mother as an illness.

A type of infanticide committed by males occurs in some wild felids and a few primate species. Males that have fought with another male and taken over territory in which there is a resident female and her litter may indiscriminately kill the kittens. Removing the newborn causes the female to come into estrus again soon. The male can then breed her, and sire the next round of offspring. Domestic tomcats have been reported to commit this type of infanticide. Thus, there is some reason to keep strange tomcats away from lactating female cats. One would expect a resident tomcat to be quite safe around kittens and, indeed, this is what some breeders have found.

HANDLING AND EARLY TREATMENT OF KITTENS

Questions sometimes arise concerning how newborn kittens should be treated by people. As many cat owners know, most mother cats are not bothered by people handling kittens for a short period of time. Contrary to what many people think, handling kittens in the first week or two of life is not detrimental. In fact, one can see signs that handling could be interpreted as beneficial to health. A few minutes of moderately stressing neonatal animals by handling has been shown to result in more rapid growth and development (see Chapter 21). As adults, the handled animals are less emotional in strange environments.

Removal of the young from the mother and nest permanently, rather than for just a few minutes each day, is quite a different matter. A limited amount of research reveals that kittens separated at 2 weeks of age are more suspicious, cautious, and aggressive as adults than kittens weaned normally (Seitz, 1959).

In raising orphaned kittens, one should first attempt to foster them onto other mother cats with kittens. Another choice would be to hand raise the kitten but to provide for maximum interaction with littermates for at least the first 4 to 6 weeks of age, since contact with littermates can compensate for lack of maternal contact. Raising an orphaned kitten singly, with no contact with other cats and minimal human contact, is the most undesirable option. In such a case, we would expect behavioral and physiological abnormalities, in a degree proportional to the duration of isolation from birth.

Section III

SCIENTIFIC BASIS FOR TECHNIQUES OF BEHAVIORAL THERAPY

In Section II we provided specific methods for treating behavioral problems. In Section III the background is provided that supports the choices for specific methods. These chapters will clarify the latitude of options and the degree of correction that can reasonably be expected with various treatments. Negative side effects are detailed more clearly here as well.

Also in this section, an overview is presented of current research findings. The interested reader can find a review of the literature with more detailed citations (Hart, 1985a).

In the first chapter of this section, we highlight the critical importance of initially selecting a pet that is appropriate for the owner's needs. This reflects our viewpoint that, as in other areas of medicine, problem prevention is often possible if informed decisions are made along the way. This involves less effort and results in less aggravation than treatment of a problem. We encourage you to support your clients in thinking about the type of pet that will work best for them.

Subsequent chapters focus on methods of modifying behavior by conditioning procedures, hormones, or psychoactive drugs. With conditioning, an animal's behavior can often be corrected without using drugs or hormones, and the same effective conditioning techniques are used for a wide range of different problems. It is well-known that hormone manipulation, as in castration, often reduces some objectionable behaviors, such as spraying by male cats or aggression in male dogs. In some cases hormone-related drugs, especially progestins, are prescribed to reduce problem behavior. In cats, changing the hormonal internal milieu provides a more direct solution to many problems, and is easier than trying to teach or train the cat to change its behavior. A number of tranquilizers are available which affect behavioral changes. While abundant data reporting the specific behavioral effects of drugs on dogs and cats are not available, some clinical results have been obtained and presently a few drugs are occasionally used to treat behavioral problems.

Chapter 21

Behavioral Aspects of Selecting and Raising Dogs and Cats

Our primary concerns in behavioral therapy are to give advice to clients regarding the best source of a pet, to determine the best method for raising it to avoid behavioral problems in the future, and how to make the pet an enjoyable member of the family. This kind of advice may be given routinely to owners of young animals when they come in for routine vaccinations, examinations, and treatment for parasites or minor health problems. The advice may often be sought over the telephone by a client even before the pet is obtained.

With an older pet, the client's behavioral complaint usually stems from unfortunate or adverse experiences in the pet's early life, and the client desires your help in analyzing the problem. In these circumstances, the animal owner may wish to follow your advice in obtaining a new dog or cat with the hope of avoiding problems of a similar nature in the future.

This chapter presents the background information that may be useful to a practitioner in presenting to a client different options with regard to breed selection, castration, and rearing practices. Many of the same general principles apply to both dogs and cats. Where appropriate, we have presented the discussion of information on the two species separately.

DOGS VERSUS CATS

In the selection of a pet dog or cat, the role of the practitioner boils down to matching the most appropriate companion animal to that person or family seeking advice. A number of parameters are considered in recommending particular pets for particular environments. The first decision is usually whether to recommend a dog or a cat. Once this basic decision is made, there are questions about size, hair length, coat color, and whether the animal should have a purebred or mixed background. Size and color are important, but we must remember that an animal is chosen primarily for its behavior (Hart and Hart, 1984), which is what contributes to the richness of the relationship it will have with people.

There is no reason to belabor the obvious differences between cats and dogs as companions. The size of cats often makes them more suitable than dogs for some environments. The fact that cats will use a litter box for elimination, as opposed to the necessity of taking a dog on a walk 3 times a day, also contributes to the popularity of cats in settings where regular access to the out-of-doors is not available.

What many people fail to recognize about behavioral differences between dogs and cats is that cats are basically asocial. Their wild ancestors were loners that lived a solitary life, with the exception of coming together for the breeding or raising of young. The domestic cat is basically the same asocial animal, although through domestic breeding we have probably selected cats with more sociable behavior. One of the reasons that cats make ideal pets for situations in which they must be left alone for much of the day is that they are not a social species. A social animal like a dog might show behavioral signs of isolation while a cat is relatively content without other animals or people around. The asocial nature of cats is also why they seem to be indifferent, if not occasionally insubordinate, to their owners.

Dogs react to people more or less as they would toward other members of the pack. They crave social interaction and love affection and attention. Dogs are social animals that readily develop dominant-subordinate relationships to human members of the family. People often make the mistake of feeling that a single cat, left home alone much of the time, will be lonely; therefore, they feel obligated to find it a companion. As many cat owners will testify, this often makes matters worse for the resident cat. The resident cat may reject the stranger, or the cats may tolerate each other for a few months and then get along miserably. Worse yet, the resident cat may start urine spraying when a new cat is brought into the house because of the extra agitation and emotional upset it creates.

SOURCE AND AGE OF ADOPTION

For the client who is interested in obtaining a 6-week-old puppy or kitten from a healthy litter, raised by an attentive mother, in a household where good nutrition and kind treatment is the rule, the only advice you can offer would be to point out the most appropriate breeds.

Some people have an opportunity to adopt an adult pet. For reasons that will become clear later, this may lead to real behavioral problems. Unless the people know the animal quite well and find its behavior appealing, they should be dissuaded from adopting an adult. A great many problems arise from adult dogs or cats that do not fit into established households. Dogs may bark incessantly. They could be impossible to housetrain. They may fight with resident pets or run away from the children. Cats may seclude themselves in nooks and crannies.

Aside from considering a particular breed, the prospective pet owner will want to select a puppy or kitten from a litter that is healthy and has been receiving adequate nutrition and medical care. For all the appeal that rescuing a neglected animal from further mistreatment may have, this type of pet is a poor risk. Inadequate nutrition and lack of kind and gentle treatment may be reflected in the behavior of the dog or cat as an adult. If the mother is obviously not adequately cared for, it is likely that the owners have also not treated the puppies or kittens very well.

Parents who are looking for a young pet for their children are often attracted to the runt of the litter. Admittedly, most runts turn out all right, but there is a greater chance of future problems involving its emotional behavior. There is a possibility of severe undernutrition early in life, since the runt is less able to compete for food. The runt is also likely to be harrassed by its littermates, and if this occurs very early in life, it may have enduring effects on its behavior. It is true that stunting of body size may not involve the brain. However, when you see marked somatic stunting there is no way of knowing how severely the central nervous system and, hence, later behavior may be affected.

In many hometown newspapers there is often a space devoted to showing the "Pet of the Week," which is available at the local animal shelter. Part of a litter of kittens or puppies is often featured in such a manner, and it is a real temptation to want to give these pets a home and rescue them from euthanasia. However, both the animal shelter and the pet store are risky places to obtain a pet because one has no way of knowing if the animals were subjected to neglect or mistreatment, the behavior of the dam, or the age the animals were weaned. As we shall see later, very early weaning has been related to the development of problems with emotional behavior.

The Dam, Sire, and Progeny

Several factors may be considered to predict the type of behavior a puppy or kitten will show as an adult. There is a good chance the behavior will resemble that of the mother, and in normal circumstances one is free to observe the interaction of the mother with members of the family, strangers, and other dogs or cats. It is also helpful to learn about the behavior of the sire, either by means of telephoning the owner or observing the sire directly. One can, in some situations, even go to the extent of "progeny testing" by inquiring as to the behavior of puppies or kittens from previous litters. This will involve some effort, but it could lead to some practical results. For example, if a client has had unfortunate experiences with a flank-sucking Doberman or a wool-chewing Siamese, it would certainly be a good idea for him to talk to the owners of the dam, sire, and siblings from previous litters.

The Age to Adopt a Pet

Many dog handlers and breeders point to a body of experimental and practical evidence to support the notion that puppies should be adopted between the ages of 6 and 8 weeks for optimal socialization with both dogs and people. There is, theoretically, a critical period extending from 3 to 12 weeks of age in which socializations or attachments in dogs are most easily made. Adoption before 6 weeks is felt to interfere with the socialization of puppies to other dogs. Adoption after 8 weeks interferes with their socialization to people.

There is little evidence of a critical socialization period for cats and, therefore, no compelling reason for adopting one before 8 weeks. As a general rule, however, 6 to 8 weeks of age would appear to be the best time to adopt a cat. This gives the new owners control over the general sensitivity of behavioral development in early life.

There are circumstances in which one may wish to leave a kitten with its mother for an extra-long time. Some Siamese cat breeders maintain that leaving kittens with the mother for as long as 12 weeks will reduce the tendency toward wool-chewing or wool-sucking. If one wishes to obtain a cat for rodent control, it would be beneficial to leave the kittens with the mother a few weeks after weaning so that the kitten may learn hunting behavior from the mother.

It is possible to obtain puppies or kittens too early. There are indications that early-weaned kittens and puppies are more suspicious and cautious than those weaned at 6 weeks (Sietz, 1959).

Orphaned Puppies and Kittens

When a mother dog or cat disappears or dies several weeks before her offspring are weaned, the offspring are deprived of important maternal-infant interactions. Regardless of the amount of cuddling and petting we may try to administer, there is no substitute for the constant interaction they would enjoy with their natural mother. Because of the laborious task of bottle feeding orphaned animals, they are sometimes split into groups for other families or people to take care of. This reduces the interactions with their littermates also. Based on experimental work with maternal deprivation, we would expect such orphans to have a tendency to be excessively cautious, fearful, or aggressive as adults. Unless one can foster orphaned puppies or kittens onto mothers with other young, or at least raise them as a litter together, one would be well-advised to consider having the animals discretely euthanized so that unsuspecting people will not be tempted to adopt them.

BEHAVIORAL CHARACTERISTICS OF DOG BREEDS

The more we know about the specific behavioral attributes or disadvantages of various breeds of dogs, the more success we will have in recommending dogs as family or personal pets. The idea of matching certain breeds of dogs with certain types of people requires a reliable, unbiased approach to determine breed characteristics. People who want dogs live in a variety of environments ranging from small apartments to houses with open space. The human environment may range from that of a single man or woman to a large family with young children. Some people who want a pet may want only a watchdog and are not concerned about the dog's behavior toward children. A family with young children may have the opposite concern. With some people, the ease of training and the absence of destructive tendencies may be of paramount importance.

Current popular literature contains statements about breed differences in behavior, such as trainability, activity level, aggressiveness, watchdog behavior, and mellowness with children. However, the perspectives obtained are from breeders and breed associations, and are likely to reflect financial and personal biases. Often the selection is approached on the basis of size and hair type. Information about specific behavioral attributes and disadvantages of various breeds of dogs is essential for the selection of a dog as a family or personal pet. What we present here is a general guideline of basic behavioral characteristics of breeds to help prospective pet owners, and professionals offering advice, in narrowing down their choices. With a general guideline in hand, one can then go to local authorities, such as dog breeders and trainers, for information about behavioral characteristics of particular lines within breeds to decide on an individual dog. Of course, to make a meaningful recommendation, one must learn something about the environment in which the dog will be living, and the personalities and lifestyle of the people involved. Most people have some idea about the type of dog they want. In many instances people will have to be cautioned to reduce their emphasis on size, head shape, or whether the dog looks cute or not, and place emphasis on the behavioral characteristics most appropriate for them.

The information we will present is from a data-based set of breed profiles obtained on 56 of the most frequently registered breeds. The information was obtained by interviewing a large number of nationally recognized obedience judges and an equally large number of small animal veterinarians. They were selected from directories so as to represent male and female informants, and to represent eastern, central, and western states, equally. There were 13 types of behavior deemed of primary interest to prospective pet owners. Behavioral characteristics were expressed specifically, in a question that asked informants to rank dogs with regard to the trait in question. The informants were also asked to compare male and female dogs in general. The list of characteristics is outlined in Table 21–1.

It was not possible for informants to talk about or promote a breed

TABLE 21–1. Ranking of Behavioral Characteristics in Order of Decreasing Reliability of Distinguishing Between Breeds

Behavioral Characteristic
1. Excitability
2. General Activity
3. Snapping at Children
4. Excessive Barking
5. Playfulness
6. Trainability
7. Watchdog Barking
8. Aggression to Dogs
9. Dominance over Owner
10. Territorial Defense
11. Affection Demand
12. Destructiveness
13. Housebreaking Ease

in which they may have had some vested personal or financial interest. The data were computer analyzed (Hart, *et al.*, 1983; Hart and Miller, 1985; Hart and Hart, 1984; 1985).

The project is predicated on three principles: (1) sufficient breed differences actually exist to discriminate many behavioral characteristics of dogs; (2) behavioral differences are known by people who have extensive experience with dog breeds and dog-owner relationships; and (3) the behavioral information that exists in the minds of dog authorities can be obtained by interviewing large numbers of different types of authorities with an interview format that minimizes the opportunity for the informants to talk about the dogs in which they may have vested personal or financial interests.

All of the behavioral characteristics listed in Table 21–1 showed significant differences among dog breeds. However, the range in F ratio reveals that some characteristics provide a more reliable basis for making distinctions among breeds than other characteristics. For example, dog breeds differ more reliably in excitability and general activity than in housebreaking or destructive behavior. All 56 breeds of dogs were ranked on each of the 13 behavioral characteristics. Thus, one can determine which breeds are the least excitable, the most demanding of affection, the best watchdogs, the easiest to housebreak, and so forth. A behavioral profile has been developed for each of the breeds that includes the dog's decile ranking in each of the 13 behavioral traits (Hart and Hart, 1985). There is not space here to present behavioral profiles on each breed, but we have summarized the information by clustering traits together on the basis of factor analyses. Each of the behavioral traits, except playfulness and destructiveness, were grouped by factor analysis into one of three groups: reactivity, aggressiveness, and trainability (Table 21–2). A statistically-based cluster analysis was then used to form 7 groups of dogs with each group being characterized according

TABLE 21–2. Grouping of Behavioral Traits Into Factors

Factor 1 REACTIVITY	Factor 2 AGGRESSIVENESS	Factor 3 TRAINABILITY
Affection Demand	Territorial Defense	Trainability
Excitability	Watchdog Barking	Housebreaking Ease
Excessive Barking	Aggression to Dogs	
Snapping at Children	Dominance over Owner	
General Activity		

to the 3 factors. This information is presented in Table 21–3. The value of these clusters is that one may look at a group of dogs which are alike in terms of their behavioral characteristics. One can then go to the specific behavioral profiles, body size, and so forth, to narrow the possibilities.

Each of the 7 clusters can be characterized by a thumbnail sketch. All clusters but one include dogs from more than one of the standard dog groups, showing that the behavioral profiles do not closely correspond

TABLE 21–3. Behavioral Profiles of Dog Breeds as Grouped by Cluster Analysis

CLUSTER 1: High Reactivity, Low Trainability, Medium Aggression		
Lhasa Apso	Boston Terrier	Weimaraner
Pomeranian	Pekingese	Irish Setter
Maltese	Beagle	Pug
Cocker Spaniel	Yorkshire Terrier	
CLUSTER 2: Very Low Reactivity, Very Low Aggression, Low Trainability		
English Bulldog	Norwegian Elkhound	Basset Hound
Old English Sheepdog	Bloodhound	
CLUSTER 3: Low Reactivity, High Aggression, Low Trainability		
Samoyed	Saint Bernard	Dalmatian
Alaskan Malamute	Afghan Hound	Great Dane
Siberian Husky	Boxer	Chow Chow
CLUSTER 4: Very High Trainability, High Reactivity, Medium Aggression		
Shetland Sheepdog	Poodle-Toy	English Springer Spaniel
Shih Tzu	Bichon Frise	Welsh Corgi
Poodle-Miniature	Poodle-Standard	
CLUSTER 5: Low Aggression, High Trainability, Low Reactivity		
Labrador Retriever	Newfoundland	Golden Retriever
Vizsla	Chesapeake Bay Retriever	Australian Shepherd
Brittany Spaniel	Keeshond	
German Shorthaired Pointer	Collie	
CLUSTER 6: Very High Aggression, Very High Trainability, Very Low Reactivity		
German Shepherd	Doberman Pinscher	Rottweiler
Akita		
CLUSTER 7: Very High Aggression, High Reactivity, Medium Trainability		
Cairn Terrier	Fox Terrier	Schnauzer-Miniature
West Highland White Terrier	Scottish Terrier	Silky Terrier
Chihuahua	Dachshund	Airedale Terrier

to the usual groupings of working, hunting, terrier, hound, and miscellaneous breeds. However, there is some tendency for overlap between the clusters and the usual grouping. A few of the clusters seem to specialize in a particular build of dog, and no cluster includes the full range of dog builds.

Cluster 1. This group of dogs, the largest cluster, exhibits high reactivity, low trainability, and medium aggression. It draws from several of the traditional dog groups. All but one of these dogs rank in the lower half on bulkiness, and the builds range from frail, as in the Maltese, to solid, as in the Weimaraner. The most prototypic breeds in this cluster are small frail dogs, Lhasa Apso, Pomeranian, and Maltese. The bulkiest breeds, Irish Setter and Weimaraner, fall at the outer edges, and could be said to have a behavioral profile that is more often found in smaller dogs.

Cluster 2. Very low in aggression and reactivity, and low in trainability, all dogs in this group are solidly built and in the upper quarter of dogs for bulkiness. Prototypic in this cluster are the Old English Sheepdog and the English Bulldog, and the smaller but strong Basset Hound falls on an outer edge.

Cluster 3. This cluster differs from the previous one in its high aggression, but it shows low reactivity and trainability like cluster 2. The Samoyed, Alaskan Malamute, and Siberian Husky are typical members of this cluster, while the Chow Chow, Great Dane, and Dalmatian fall at the extremes of the cluster. All of these dogs average at least 21 inches at the withers and are in the upper half of dogs for bulkiness.

Cluster 4. Very high trainability, high reactivity, and medium aggression characterize the dogs in this cluster. The dogs range in bulkiness from frail to middle, 9 to 20 inches high. The Shetland Sheepdog, Shih Tzu, and Miniature Poodle are most typical of this diverse cluster.

Cluster 5. Low aggression and reactivity, and high trainability characterize this cluster of dogs that ranges in bulkiness from light to the most powerful. This cluster is most typified by the Collie, Vizsla, and Golden Retriever. The Australian Shepherd and Brittany Spaniel are on the outer edges of this cluster.

Cluster 6. This small group of sturdy, strong dogs exhibits very high aggression and trainability, and very low reactivity. The Akita, German Shepherd, Rottweiler, and Doberman are the only dogs in this group.

Cluster 7. Very high aggession, high reactivity, and medium trainability are manifested by these dogs that range from fragile to middle bulkiness. The West Highland White Terrier, Silky Terrier, and Chihuahua are near the center of this cluster, and the Dachshund, Airedale Terrier, and Miniature Schnauzer are on the outer edges.

The guideline presented here by means of the cluster analysis is, of course, just a general guideline as to the suitability of certain breeds for various environments or human-animal interactions. Early experience and the behavior of the pet owner play an important role in the behavioral characteristics of any particular dog. Obviously, these inputs into the behavioral profile of an individual can be manipulated.

If the selection of a breed can be narrowed down to a single breed, then one could go to particular dog breeders or trainers for more information about behavioral characteristics of particular lines within the breed to decide on an individual dog. One might also think at this point about applying one of the tests of puppy temperament to decide on a particular dog within a litter. A major problem with puppy tests is that the validity of the tests, that is, the predictability of puppy tests for adult behavior, has not been established.

A word about mongrels versus purebred dogs: Many or most mongrels turn out to be very acceptable pets. The advantage of selecting a purebred over a mixed breed is that with the purebred one has more success in predicting what the dog will be like as an adult.

GENDER AND CASTRATION CONSIDERATIONS IN DOGS

Some of the same behavioral traits that vary from breed to breed also vary significantly between sexes. In Figure 21–1, we present a listing of traits in which males rank higher than females and traits in which females rank higher than males (Hart and Hart, 1985). Since these are the same traits for which breeds were ranked, one can see the interactions here. If one wishes to obtain a dog from a cluster that ranks as moderate with regard to aggressive traits, but wants to minimize the aggressive tendencies, one could choose a female rather than a male. On the other hand, if one were choosing a dog from a cluster in which aggressiveness ranks low, the choice of a male versus a female would not be such an important consideration.

Since male dogs are occasionally castrated for behavioral reasons, the question arises as to the degree to which the traits that males display more than females may be altered by castration. Too many times people feel that castrating a male dog will make it calmer, less destructive, less aggressive toward the owner, or better with children. Roaming, mounting of other dogs or people, urine marking in the house, and aggression directed toward other male dogs are traits most likely to be reduced by castration according to clinical documentation. These behavioral characteristics, which quite obviously differ between males and females, are most amenable to castration. The tendency to be dominant over the owner, or to display more general activity than females, may also be altered by castration, but no clinical or experimental observations have been presented to document such an effect. With behavioral characteristics that one expects to be amenable to castration, clinical data has revealed that when dogs are castrated as adults in an attempt to alter these behavioral patterns, the effectivenss is in the range of only 50 to 60% (Hopkins, *et al.*, 1976). Although clinical data are not available, one

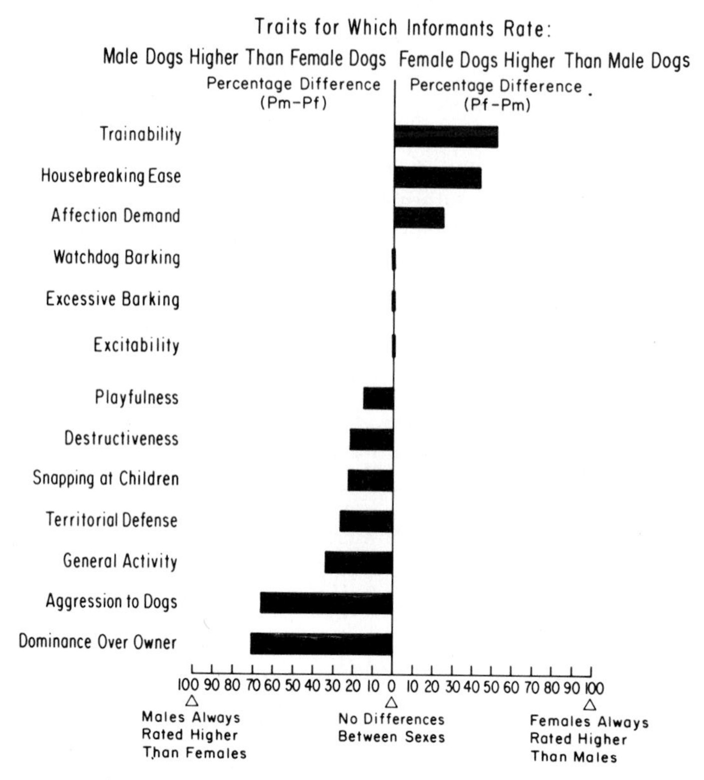

Fig. 21–1. According to the ratings of experts, male dogs score higher on some traits, females on others. On three traits there were no significant differences between males and females (From Hart and Hart, 1985; reprinted with permission, American Veterinary Medical Association).

would expect that prepubertal castration in male dogs would not be any more effective in preventing the occurrence of these behavioral patterns than postpubertal castration in eliminating them once the behavior has been displayed.

BEHAVIORAL CHARACTERISTICS OF CAT BREEDS

What breed of cat is best with children? What would be the most appropriate breed for a single person who wants minimal interaction with a cat? These are the kinds of questions that anyone recommending a cat could be confronted with. As with dogs, one must be wary of trusting the trade books which deal with particular breeds because either they mention very little about behavior that can be counted on, or they point out only the desirable behavioral patterns of the different breeds. Most people working with cats have some impressions about behavioral characteristics and these impressions are usually based on limited personal experience which may or may not be typical for that breed.

Breed-specific Behavior of Cats

The information presented here is from an informal survey of cat show judges who have a wealth of experience with various breeds (Hart, 1979a; 1980a). Since judges converse frequently with breeders, talk to other judges about breeds, and have raised cats of several breeds themselves, they are undoubtedly the best source of information. With information about breed-specific behavior, one might think of recommending a particular breed for certain people or situations. One might recommend a Siamese or a Burmese cat for a family with children because they interact more readily with family members than cats from other breeds. The following are behavioral characteristics obtained from judges registered with the Cat Fanciers' Association or the American Cat Fanciers' Association, the largest and second largest cat registry associations, respectively. The judges emphasized that there are major individual differences in behavior between cats of the same breed, so the generalizations are not going to hold for all cats of that breed. An Abyssinian may act like a Burmese and vice versa.

The value of selecting a purebred rather than a mixed breed is that one has more success in predicting what a cat will be like as an adult. Because the environment has a more pronounced affect on behavior than morphology, prediction of behavior is less successful than prediction of body shape or coat color. The method of assessing behavior in this study is not data-based or quantitative in the same sense as the study on dogs. Instead, the informants were used for narrative responses to obtain nonstatistical consensus statements.

Siamese. This most popular of all breeds originated in Thailand and is reported to be the most outgoing with strangers and demanding of affection and attention. Some informants feel this reflects a desire for warmth because of their light coat. Although Siamese do not display the one-person-cat behavior that we associate with more fearful cats, they become strongly attached to their owners and recognize their owner's voice at a distance. Siamese cats extensively vocalize in a style that many people refer to as "talking." They easily vary the pitch of their meows, especially if this gets a reaction from the owners. The vocalizations are objectionable to some people, especially in small apartments or when females come into estrus. Loving to be held, snuggled, and carried about, the Siamese is a good pet for gentle children.

Persian. The Persian's beautiful long fur coat attracts many people, but it requires combing and brushing several times a week. Behaviorally, these cats are reported to be somewhat lethargic, reserved, inactive, and do not seek affection as do the Siamese. Seemingly not desiring close contact, they are more comfortable with being petted as they lie on the floor. The aversion that Persians have to being held is partially attributed to their tendency to get too hot on a person's lap, due to their heavy coats. The judges interviewed rated the Persian as the safest cat if one wanted to avoid a song bird predator. This could be related to the low activity level.

Burmese. Like their close Siamese relatives, the Burmese are reported to be demanding of affection. They are considered good family pets, if the family members want a cat that is affectionate, easygoing, and playful. Compared with the Siamese, they vocalize less and are less outgoing to strangers, but are not as withdrawn as some other breeds. A former reputation of the Burmese breed for being temperamental is apparently no longer appropriate. This breed is considered suitable for a family with active (but gentle) children.

Abyssinian. This breed is the most feral looking of the popular breeds and is reportedly usually more shy and fearful of strangers. The Abyssinian may be too nervous to make a good cat for children. Apparently it does not like snuggling as a lap cat, but enjoys being petted at a distance.

Manx. People are attracted to this breed by its overall roundness: a rounded head, arched back, round rump, and lack of a tail. The behavior of the Manx seems to be more variable than in other breeds, since the judges did not agree in assessing this breed. Although some classified the cat as withdrawn and very wary of strangers, others felt the cat was quite playful and would be good with a family.

Himalayan. Rapidly growing in popularity, this attractive cat has the heavy, short, well-rounded body and long fur of a Persian and the coloring of a Siamese, and its behavior is apparently intermediate between the two breeds. The Himalayan is not as outgoing to strangers as the Siamese but is not as reserved as the Persian. With its heavy coat, the Himalayan will tend to become too warm and thus not appreciate extensive cuddling.

Russian Blue. While this breed's reportedly shy and withdrawn behavior is not appealing, the dense, plush fur coat feels like a short-coated beaver. Trace a design in the coat with your finger and it remains. Being one of the most shy and withdrawn of all breeds, it will avoid your guests and other cats.

Rex. People that are allergic to cats are often able to tolerate the Rex, which has no fur undercoat. However, the jittery behavior of this breed makes it difficult to recommend; it is easily upset, withdrawn, apprehensive, unpredictable, actively high-strung, and hard to handle.

GENDER AND CASTRATION CONSIDERATIONS IN CATS

If a cat has a behavior problem that becomes intolerable to the owner, it is not going to be much of a companion. The single most important behavioral problem in cats is urine spraying. In fact, because of the seriousness of the behavioral problem, the animal may have to be disposed of if therapeutic measures are not successful. Fighting with other cats is the next most serious problem, and this behavior can also be very disconcerting to owners. By carefully selecting an animal, the likelihood of having a cat that engages in these problem behaviors can be minimized. It is common practice for male cats to be castrated before puberty, and the assumption is that prepubertal castration is more effective in preventing objectionable urine spraying, fighting, and roaming than postpubertal castration is in eliminating these behaviors once they have begun. However, urine spraying is common in prepubertally castrated males and spayed females, and the administration of progestins to control this behavior has become routine in feline practice (Hart, 1979d; 1980c). It is not unusual to find male or female cats who had been neutered at 6 months of age and then began spraying as late as 3 or 4 years of age. The onset of spraying is often related to the owners who introduce new cats into a household with other cats, change households, or alter a major aspect of the cat's lifestyle, such as converting an outdoor cat into an indoor one.

A clinical survey to obtain reliable information on urine spraying used tabulated questionnaires of cat owners to assess behavior (Hart and Cooper, 1984). The survey revealed that the probability of frequent urine spraying by prepubertally castrated male cats is very close to 10%, and the probability of frequent spraying by prepubertally spayed females is 5%. Thus, females are generally a safer bet than males for someone who needs to avoid this problem. Male cats are also more likely to spray when they are living in a household with another female cat. (see Chapter 14). What this means is that if a person already has one resident cat and wants to get another, but also wants to diminish the possibility that either the new or resident cat will take up spraying, the second cat should be of the same sex.

RAISING KITTENS AND PUPPIES: THE EFFECTS OF EARLY EXPERIENCE

Of all the domestic animals kept by man, our relationship with dogs and cats is unique. These are the animals we bring into our homes and raise as members of the family, often from the day they are born. Our access to young kittens and puppies and the ability to manipulate their environment in early life gives us great power over the development of their behavior. The effects of early experience translate into a number of conceptual issues with regard to raising puppies and kittens. In this section we will examine the three different periods within the early life of dogs and cats that have special significance in terms of behavior. Certain principles and information that have very significant implications emerge with regard to each of these periods.

Prenatal and Neonatal Life

In the first two weeks after birth, and even extending back into the late prenatal period, the brain is undergoing rapid development. Neurons are still multiplying to some extent and there is extensive growth of axons and dendritic branches. Enzyme mechanisms in neurons are changing and there are a multitude of changes in the neuroglial cells which have a supportive and nurturing role for the neurons. Once neuronal growth and development have occurred in early life, there is little alteration or growth of brain constituents later. For example, brain cells stop multiplying after the first few days of life. While axon and dendritic growth can occur in the adult animal, this is relatively unimportant. Thus, certain experiences that influence the brain just after birth, particularly those which are accompanied by changes in hormones in the developing organism, can have effects which are in essence "locked in" to the developing brain. These effects may be expressed in terms of behavioral changes, since behavior is a function of brain activity, or they may be reflected in terms of certain physiological responses. The effects can tend to be irreversible.

One of the examples of such an effect which has been studied relatively intensely is the influence of early malnutrition on brain growth and subsequent behavior. A severe protein deficiency, as low as one third to one half of the normal level, can impair normal brain development in the neonatal period, resulting in stunting of brain growth. These effects have been studied experimentally in laboratory rodents, but the results obviously apply to dogs and cats that are also in a relatively immature state when they are born. Rather than having pronounced effects on learning ability or learning behavior as one might expect, laboratory experiments point to a relatively permanent change in the emotional behavior of animals so affected by early malnutrition. These emotional changes are exemplified by hyperemotionality around the time of feeding with profuse excitement and intense orientation toward feeding.

About the only area in which the small animal clinician is likely to be concerned about effects of malnutrition on brain growth is with excessively large litters. The smaller animals of the litters, particularly the litter runts, are most likely to suffer from such a nutritional deficiency. The strongest animals in the litter may control the nipples most of the time by virtue of their size and strength. This exacerbates the problem of the already short milk supply. Some owners noticing what is going on may attempt to supplement the food for the litter runt, but this could have the effect of simply allowing the animal to survive on a very marginal basis where it would otherwise die.

Neonatal Handling and Stress

Whether an animal owner is a regular breeder of dogs or cats, or is a person who just decides to have one litter, there are always concerns and questions about the dangers of handling the newborn puppies or kittens. Many feel that the kittens and puppies should not be handled for at least the first couple of weeks. In families where children are present at the time of the birth of the young animals, this can be an almost impossible assignment. What kind of advice should we offer about the effects of such early handling?

From work with rats and mice, and to a more limited extent with rabbits and cats, the concept has emerged that the handling of neonatal animals for a few minutes each day results in more rapid development of many organ systems when compared to nonhandled control animals. The effects typically reported are accelerated maturation of the central nervous system, earlier eye opening, earlier development of motor coordination, earlier onset of hair growth, and accelerated weight gain. The most important behavioral effect is that when the handled animals are compared with their nonhandled controls in a situation which tends to evoke emotionality and freezing, such as placing the animal in a completely strange environment, the handled animals appear less emotional and intimidated by the strange environment. These behavioral observations are backed up by signs of less autonomic activation as well. One experiment which tends to confirm these observations in cats reported that when kittens were handled for 10 minutes a day starting just after birth, their eyes opened a day or so sooner than nonhandled controls and they developed the characteristic coloration slightly earlier. The handled kittens emerged from the nest about 3 days sooner and, in general, were more active than controls (Meier, 1961).

The handling effect can also be achieved by administering a mild electrical shock to the neonatal animals, as well as exposing them briefly to cold temperatures, or even shaking them up in a shoe box. In other words, the handling appears to be stressful and any type of stress achieves the same effect.

One of the leading investigators in this area (Levine, 1969) theorizes that handling or other types of stress lead to optimal development of the pituitary-adrenocortical system, so that in the adult the ouput of adrenocortical steroids more appropriately matches the environmental stresses. For example, if the handled animal as an adult is exposed to maximal stress, the condition is met with maximal production of adrenocorticotropic hormone and cortical steroids; more moderate stresses are met with a more moderate output of the hormones. On the other hand, it is proposed that the nonhandled animals' cortical steroid output tends to be maximum under minimal stress as well as maximal stress, and is thus less adaptive or suited to the animals' needs.

Whether the accelerated growth and reduction of emotionality would necessarily be advantageous to the development of animals in the wild is debatable. It has been argued that the handling effect is a function of artificial and laboratory rearing with little relevance to wild animals (Daly, 1973).

It is probably fair to say that a moderate amount of handling of newborn kittens and puppies is not going to be detrimental to their physiological and behavioral development. There is an indication that there may be some advantages of handling in terms of enhanced development and less emotionality of the animals as juveniles and adults, which are useful for animals, at least on the domestic scene. Therefore, clients could be advised that some handling is permissible but that parents and older individuals in the family should watch to make certain that this does not become excessive.

In contrast to the rather extensive amount of experimental work conducted with neonatal animals, there has been very little done with animals prenatally. Admittedly, animals in this state are generally isolated from the world in terms of tactile, visual, auditory, and olfactory sensations. However, the nervous system is still sensitive to hormones and drugs, particularly those that may come to it via the placenta from the maternal blood supply. We know, for example, that androgens such as testosterone, and also androgens metabolically derived from progestational compounds given to mothers, can affect the developing nervous system of female fetuses and result in masculinizing effects on behavior. This topic will be dealt with in Chapter 23. One experiment on rats is worth mentioning here. Female rats made continually fearful in late gestation were found to bear young which, when tested as adults themselves, were more emotional than the control animals. This effect was attributed to the repeated arousal of high levels of epinephrine in the mother and the possible effect that this hormone has on the developing brain of the offspring through the placenta. If the results of this experiment are to be taken seriously with regard to dogs and cats, then one would want to discourage subjecting bitches or queens in late stages of pregnancy to repeated fear-inducing circumstances such as shipping across the country or boarding in a kennel. Of course, this is generally good advice regardless of the effect on the offspring's behavior.

MATERNAL AND PEER INFLUENCES

Maternal Deprivation

An aspect of maternal behavior that is obvious to anyone that has raised kittens or puppies is that the mother spends an extraordinarily high percentage of her time with the infants. The time spent with the offspring is divided between suckling the young, grooming the head, body, and anogenital region, retrieving, and sleeping with the young in close contact. A fairly extensive body of literature on laboratory animals indicates that the severe disruption of this relationship can have important consequences on the behavior of the young even when they reach the juvenile period and adulthood. Understanding the effects of deprivation is important because practitioners are frequently called on to give nutritional and medical advice to individuals who wish to raise animals that have either been weaned very early, or left as orphans because of the death of the mother. It is unfortunate that, in the literature available to the lay public and veterinarians, virtually nothing is mentioned about the possible adverse effects of raising orphaned animals, or of ways to minimize them. This gap occurs even though the veterinary literature specifies precautions for raising orphaned animals in terms of popular formulas for milk substitutes, temperature control, and even recommendations for preventing infants from nursing on each other. Accurate observations on the behavior of dogs and cats weaned at a very early age are rare. However, one reliable study has revealed that kittens separated from their mothers at about 2 weeks of age, even though they are left together with other littermates, are reported to be more suspicious and cautious than those weaned at the usual time of 6 to 8 weeks (Seitz, 1959). These observations on cats are confirmed by more detailed studies on rats showing disruptions of normal emotional responses as a function of very early weaning. These adverse effects of early weaning are not apparent when kittens are weaned at four weeks of age, which indicates that separation from the mother must occur at a very early age to produce undesirable behavior.

A phenomenon related to maternal deprivation is the finding that animals raised in excessively large litters may be more emotional in strange areas than those from smaller litters. This effect may result from the reduced maternal attention that is available for each of the offspring.

One effect of early weaning commonly seen in dogs and cats is an increased occurrence of non-nutritional sucking. Kittens or puppies that suck excessively on each other's ears or penises can cause irritation of these body parts.

Although it has not been studied, there is reason to wonder if the wool chewing displayed by some Siamese cats is a function of the time of weaning. Some breeders have reported that they were able to diminish this tendency in their line of cats by allowing the kittens to stay on the mother for longer than the normal weaning time. Since wool chewing or wool sucking is a serious problem at times, it may be worth exploring in some detail.

Littermate Interaction

As soon as the eyes and ears begin to function in kittens and puppies, the surrounding environment takes on new importance. It is only logical that external stimuli and initial experiences at this time have a great impact on the development of a dog or cat. Under normal circumstances the most important external stimulus for animals of this age outside of the mother is littermates. These littermates are not only a source of playful interaction but also of warmth and comfort. In the latter sense the littermate interaction can substitute for maternal care. During playful interactions innate social responses are developed, refined, and given limits (Dunbar, 1979). Cats learn to interpret threats from each other and when to run and when to attack. In dogs these social interactions are even more complex and more important. The roles of being dominant or subordinate are developed and refined. Aspects of sexual behavior are also developed during playful interactions among puppies.

Some experiments have shown that puppies raised at an early age isolated from littermates and other dogs tend to be much more submissive and less competitive than their socially raised controls. Abnormal sexual responses have also been one of the consistent findings of dogs raised in isolation. During tests of sexual behavior in adults, such male dogs played excessively and persistently mounted in an inappropriate and disoriented fashion. Female dogs raised in isolation do not show the disruption of female receptivity that is analogous to the disruption of male behavior in the males. Incidentally, male cats do not seem to show this disruption of sexual behavior either.

Clinical Implications

Several points can be made from the foregoing discussion regarding maternal and peer deprivation. First of all, one could make the point that to invest a great deal of time and energy in raising puppies or kittens orphaned at a very early age may not be worth it. Certainly there is no evidence that such animals will turn out to be any more behaviorally desirable than those raised in a more normal fashion, despite the loving care given to them by their human handlers. Undoubtedly, human comfort, handling, and petting can partly substitute for maternal presence, but it is obviously impossible for a person to substitute for the round-the-clock care that a mother cat or dog is able to offer her offspring. One probably owes it to clients who are thinking of adopting an animal to avoid providing them with a dog or cat that has been raised as an orphan. While we would not expect most of these animals to show undesirable behavior, it is fair to say that such animals have a higher chance of turning out to be undesirable, particularly with regard to emotional behavior, than normally raised animals. People should not adopt a dog or a cat from a litter much earlier than six weeks of age, regardless of a child's impending birthday or the desire to have the animal as a Christmas present.

Puppies and kittens that are orphaned and hand-raised should, if possible, be kept together rather than separated by dividers in an incubation box or separated among several households. Peer interaction can counteract some of the undesirable aspects of maternal deprivation. By the same token, raising an orphaned cat or dog when it is the only animal in the litter is riskier than raising three or four such orphaned animals together. Finally, if young animals are orphaned or weaned at about three weeks of age, raising them separately without interaction with littermates or other animals is likely to create more problems with dogs than with cats. Since cats are basically asocial, and often kept as the owner's only pet, they are not susceptible to the socialization effects that are important for dogs. The period of three to six or eight weeks of age, when puppies commonly interact with each other as littermates, tends to be important in helping to refine and develop the social responses that are so important for dogs getting along with other dogs and even people later as adults.

In the foregoing section some comments were made about the possibilities of the litter runt being subjected to the effects of early undernutrition, particularly when it is a member of an excessively large litter. One potential problem of the litter runt in littermate interactions is that the animal is likely to be harassed socially much more than other members of the litter. The continual harassment and possible maternal neglect (there is no particular reason to feel that the mother will take pity on the runt) would imply that clients should be advised not to adopt the litter runt as a personal pet.

Critical Periods of Attachment and Socalization

The weeks surrounding the weaning period, and the age when puppies and kittens are adopted by new owners, comprise a period when the animal becomes attached to people or animals or other species, and when it acquires a number of social responses that contribute to its desirability as a pet. This is a time when neglect or mistreatment may have profound effects on the animal's behavior as an adult. We can take advantage of this period to induce certain social responses that may not normally be made by the animals. The discussion of these effects is separated into two categories—those of general effects and those specifically related to the socialization of dogs.

Early Social Exposure

When an animal first starts to interact socially with members of its own species, or other species, including humans, these experiences will have profound and lasting influences. Cats are asocial by nature, and if raised in the wild, will interact in a limited way with other cats, dogs, or people. Cats, however, if exposed to dogs early in life, will often continue to be friendly and fearless of dogs, or at least the individual dogs the cat is familiar with. In fact, about the only circumstance in which one sees a cat that gets along well with a dog is when a cat has been raised as a kitten to get along well with a particular family dog. The same goes for the reaction of cats to children. Those cats who seem to do well in social interactions with other cats, and are the most tolerant of living in a household with other cats, are also those that have been exposed to other adult cats when growing up and that continue to live in a household with several cats. A cat that is adopted into a home with no other dogs or cats soon after weaning and is then exposed to dogs or other cats only as an adult, frequently reacts in a very aggressive or withdrawn manner when placed in a household with these animals.

The same general effects of early exposure apply to dogs. Many of us are acquainted with dogs that seem to be extremely fearful or insecure around children, even to the point of acting aggressively toward children out of fear. In many of these instances, this type of fear can be traced back to the animal's early history in which children were not introduced to the dog on a regular basis as a puppy. There is such a big perceptual difference between human adults and children for some dogs that apparently the development of social responses to adults does not generalize over to children.

These early socialization effects in dogs and cats also apply to exposure to certain types of environments. Those dogs that travel best in the car or on airlines are those that are exposed to these experiences early in life.

Socialization in Dogs

Dogs are by nature social and in most instances they have a well-developed, inherited basis for displaying specific social responses. Behavioral patterns involved in threatening and fighting are basically innate. Dogs are also capable of displaying submissive postures or gestures such as putting the head down, diverting the eye glance, dribbling urine, and even rolling over on their back. These submissive gestures have adaptive value; once they are perceived by the opponent, the aggressive attack is inhibited. The subordinate dog is spared serious injury without fleeing or leaving the group. The dominant dog gets its way and even avoids being injured itself in a fight.

These social responses apparently undergo some refinement and development in dogs that require interaction with littermates. Furthermore, some social responses are applied toward human handlers, presumably because dogs frequently react to their human handlers and owners as though the human was a member of the dog pack. Hence, a dog learns to be subordinate to people, basically reacting to the physical basis of a threat or hitting.

Scott and Fuller (1965) have emphasized this socialization concept, referring to it as a critical-period hypothesis. They suggest that the period between three and twelve weeks of age is one in which the development of social responses is particularly important. If a dog is deprived of other canine contact from about three weeks of age on, it is likely to be socially inept. For example, it may not emit the correct social signals indicating submission for the purpose of turning off an aggressive attack by another dog, or it may not understand the submissive gestures of another dog and arrest its own attack. Some dogs that persistently get into fights with other dogs and seriously injure themselves or the other dogs may fall into this category. In other instances a dog may act extremely withdrawn or submissive as a function of inadequate exposure to other dogs. The importance of a dog shaping these social responses for interaction with people is also emphasized. If people do not interact much with a dog until after it is twelve or fourteen weeks of age, the dog may either be excessively fearful of people or it may be uncontrollably aggressive toward people because it has not developed a repertoire of subordinate responses.

Therefore, when a person is going to adopt a puppy into a family, particularly where there are no other dogs or puppies around, it is wise not to adopt the dog until approximately six to eight weeks of age so as to let the dog develop a full set of social responses, at least to its littermates. Of course exposure to other dogs after adoption is also desirable. In the case of dogs raised as puppies in a kennel where there may not be frequent human contact, they should be adopted no later than eight weeks of age to take advantage of the latter part of the socialization period for the shaping of desirable social responses toward people. Of critical importance here is that the dog learns to take a subordinate role to all people it contacts, including children in the family. Early enrollment in an obedience class is valuable.

Experimental observations on dogs have also suggested that excessive physical mistreatment during the socialization period may have the effect of making a dog overly submissive and withdrawn around people.

An important concept gained from understanding the socialization hypothesis is how a specific recommendation might be made for a specific household. For example, if a couple desires to have a dog that will respond favorably by not being fearful while still subordinate to children, but they have no children at the present time, then you might recommend that they look for a puppy that is born to a family breeder that has children with which the dogs frequently interact. They should consider adopting the dog somewhat late in the socialization phase, perhaps as late as twelve weeks of age, so as to insure favorable behavior toward children. Following adoption, the couple might regularly have the dog exposed to children.

There are other circumstances in which a potential dog owner may wish to adopt a dog particularly early in the socialization phase. This might be an instance where a certain breed or line of dog is desired, but the potential owner is concerned about the family environment into which the dog is born. They may also find something objectionable about the behavior of the bitch and might be concerned that some of the bitch's emotional responses might tend to be acquired by the puppy.

CONCLUSIONS

Cats and dogs are the only species of animals that are regularly subjected to the same kind of human treatment from the day of birth as human infants. We are all aware of the profound influences that the early experiences of children have on their behavior as adults. Too many animal owners do not recognize the adverse effects of isolation and neglect on young animals, overlooking that they are sensitive, developing organisms during the socialization period. An understanding of the concepts presented in this chapter is useful in offering clients advice in adopting and raising animals. Choosing the most appropriate breed is only part of the issue. There is a pronounced interaction of early experience with genetically acquired attributes. Once the pet is selected, the genetic aspect is fixed and what we have left to manipulate to the pet's and our advantage is environment and experience. We have alluded to instances in which a practitioner may recommend particularly early adoption of a dog or cat and those circumstances in which later adoption might be recommended. When someone seeks your advice regarding adoption and raising procedures, it is advisable to investigate, with some degree of detail, the particular environment that the animal is destined for, and the type of behavior that the individual expects of the animal. A dog that will be expected to perform in the show ring might be handled differently than the dog that is expected to romp around with children.

Some of the facts and concepts mentioned here have a direct bearing on the question of the source from which an animal should be obtained. It is obvious that the best source is a healthy litter, raised by an attentive mother, in a household where good nutrition and kind treatment are the rule. Not only is this environment most favorable from the standpoint of early experience, but one has the opportunity to observe the behavior of the mother.

Experimental work on animals treated differently during the neonatal period indicates that some degree of stress is not detrimental, and may possibly be beneficial to animals that are born relatively immature, such as dogs and cats. This is probably the ideal time to trim dewclaws or dock the tails from those breeds in which this is required. The stress is least likely to have any lasting effects on the animal's behavior when it is conducted early. In terms of the handling of kittens and puppies by children in the family, there is no experimental reason to deprive children of this pleasure, but it should be done in moderation and with soft and kind handling.

Chapter 22

General Use of Conditioning Procedures

Conditioning procedures provide the most important tool in solving most behavioral problems. Throughout Section II, which discusses the treatment of specific problem behaviors, references were made to a number of conditioning procedures. The purpose of this chapter is to discuss these procedures in a unified way and to give a theoretical basis for their use. There are, of course, other techniques for solving specific behavioral problems, such as the use of drugs, or taking advantage of species-typical behavioral characteristics. But the technique of conditioning is by far the most widely applied. Even when some of the physiological approaches such as castration or progestin therapy are employed, the concurrent use of conditioning procedures is also often indicated.

The time required for an office consultation to discuss conditioning procedures with a client ranges from 15 to 60 minutes. Some of the procedures require a considerable time commitment from the client and daily practice sessions. Remote punishment requires some preparation time, but usually no daily practice trials. Before spending the time needed to discuss the details of a conditioning program, it is necessary to determine the extent to which the client is interested in and able to stay with its demands.

When confronted with a behavioral problem, it will be necessary for the therapist to select the appropriate procedure and then determine how this program can be shaped for the specific case. This usually involves some kind of diagnosis regarding the cause of the problem. The diagnosis will automatically suggest one or more therapeutic approaches. A useful tip to keep in mind is that clients will usually have tried some sort of punishment or behavior modification before seeking your advice. Since your advice is being sought, their attempts have obviously not worked. Therefore, think along the lines of trying something that is the opposite of, or quite different from, what the client has tried. If the client has tried interactive punishment, think of using remote punishment or counter conditioning techniques.

An oversimplified first-level approach to dealing with behavioral problems is to think of two possibilities: punishing the behavior that is undesirable, or rewarding good behavior. This requires an investigation into the circumstances in which an animal is good, as well as determining the circumstances in which its behavior is bad. If a dog is being destructive around the house or yard, it is important to know when it is being good around the house and yard. The use of remote punishment, if it works, is an expedient way of dealing with problem behavior. If remote punishment is not indicated, or has been tried and proven unsuccessful, the only alternative is to eliminate the problem behavior by one of several conditioning techniques, which usually involve enhancing and rewarding desirable behavior. The conditioning techniques include systematic desensitization, to eliminate learned phobias or habituate innate phobias; counterconditioning, to establish a new response that is incompatible with the performance of the undesirable behavior; affection withdrawal, to get an animal to approach and obey people that were previously disliked; and extinction, to eliminate some objectionable behaviors that were acquired through prior learning such as attention-getting behavior.

All of these techniques are best understood if they are related to some of the basic concepts involved in conditioning processes. In this chapter three types of conditioning procedures are discussed: habituation, classical conditioning, and operant conditioning. Following this, the techniques that have been derived from these conditioning processes—systematic desensitization, counterconditioning, extinction, and affection withdrawal—are discussed as they might be applied to specific problems. A few examples will be mentioned to illustrate both the conditioning procedures and the therapeutic techniques, but specific details for the employment of these techniques are mentioned in Section II, which deals with specific behavioral problems.

CONDITIONING PROCESSES

Habituation

Every animal has a set of behavioral patterns that is innate and that serves to protect it. These include startle responses to any sudden unfamiliar stimulus, fear or anxiety to new or strange objects, and anxiety from being isolated or left alone. These responses are genetically programmed and protect an animal from surprise attacks from enemies, predators, and the like.

Every environment produces strange and startling stimuli, but animals have the ability to habituate such stimuli that are repeated over and over again, especially if they are never followed by any harmful or painful stimuli. When the animal habituates to a stimulus, it is to that specific stimulus, and not to everything that might cause a startle response. A dog that habituates to airplanes going overhead may not habituate to the sound of firecrackers. The paradigm for habituation may be given as follows:

Unconditioned Stimulus (sharp noise) – – – – → Emotional Activation

Unconditioned Stimulus—Repetitive – – – – → Weak Emotional Activation

Unconditioned Stimulus—Repetitive – – – – – – – – → No Response

Habituating to harmless, repetitive stimuli that initially cause a startle reaction is very adaptive. It saves an animal from twitching all day in response to every sound produced by a breeze.

Young animals habituate more readily than adult animals. Hence, it is suggested that situations that evoke fear or anxiety in dogs and cats, such as being taken in an automobile, walked through heavy traffic, loaded into airline shipping crates, or having guns shot around them, be performed when animals are very young. Hunting dogs are usually habituated to gunshots when they are puppies by having a starting pistol frequently fired around them. Habituating a dog to gunshots when it is an adult is a more laborious and lengthy process. Cats have a natural fear response to interactions with dogs, but kittens may be habituated to rough-and-tumble interactions with puppies at a very young age. It is almost impossible to habituate an adult cat to this type of interaction with dogs.

Habituation may be accomplished by repeatedly presenting a stimulus that causes the fear or startle response. This is sometimes called flooding. This type of habituation works well with young animals. Generally pet dogs and cats are habituated, by virtue of flooding, to the ordinary stimuli that are in the urban environment, such as automobile rides, machinery noises, sounds of a vacuum cleaner, power tool noises, and the like. Dogs and cats that grow up in areas where there are frequent thunderstorms usually habituate to these sounds.

Habituation is an active process, and for animals to continue to be habituated, they must be continually exposed to the stimuli, at least on an occasional basis. A puppy that has been habituated to gunshots may regain the innate fear of gunshot sounds if it does not hear any gunshots for a couple of years. In other words, habituation is an active process that requires maintenance.

In the clinical setting we are confronted with animals that have unhabituated fear or anxiety reactions. These may exist because the animal was never habituated to the stimulus that causes the reaction, or perhaps it was once habituated, but since the stimulus was not continued on a long-term basis, the animal lost the habituation. In adult animals unhabituated fear reactions are often much more intense or severe than in young animals. Dogs may throw themselves through sliding glass doors at the crack of a thunderbolt or on hearing firecrackers if they are not habituated to the stimuli. Usually the animal attempts to be where the owner is, and if the owner is outside and the dog inside, the animal will do anything it can to get outside, and vice versa.

Unhabituated anxiety reactions in adult dogs are usually enhanced by classical or operant conditioning. A dog that gets hurt when it reacts violently to the sound of a thunderstorm is basically proving to itself that thunderstorms are harmful, and this enhances the fear reaction the next time. Unhabituated fears may also be enhanced by the owner's attempting to comfort the animal when it shows signs of fear. This comforting is rewarding to the animal, so it learns that the fear which comes naturally also pays off in terms of extra comfort and affection. Thus, in dealing with unhabituated fear and anxiety reactions in adult animals, you usually have to examine the conditioning that may have intensified the reaction.

The therapeutic approach to unhabituated fear and anxiety reactions in adult animals is gradual habituation. If we present the stimulus that evokes the reaction at full force repeatedly (flooding), as we might with a puppy, the emotional state produced may be so intense that it is aversive itself and reinforces the fear reaction. This can prevent the reaction from being habituated. If we present the stimuli that produced the fear reaction in a mild fashion, and this is done repeatedly, then the fear or anxiety can be habituated at that mild level. Once this is accomplished, we can increase the intensity of the stimulus and repeat it again, accomplishing habituation at the new level. Over a series of stages the stimulus intensity is gradually increased until we have, in fact, habituated the animal to the intensity at full strength. This process is called systematic desensitization. The technique of systematic desensitization is used with unhabituated fears and with fears or phobias that may be acquired by the process of classical conditioning. The manner in which a stimulus is initially presented in a mild form, and later gradually increased, varies. For loud sounds there is a gradient of intensity: for separation anxiety, it is the duration of time the animal is left alone; for fear of strangers, it is the distance to the strangers; and for submissive urination, it is the degree of excitement involved in the greeting reaction and the intensity of the person interacting with the animal.

Classical Conditioning

The two ways by which animals may acquire new responses are through classical conditioning and operant conditioning. In classical conditioning we are dealing with responses that are reflexlike and involve contraction of smooth muscles and secretion of the glands. These responses are innately evoked by specific stimuli. Conditioning occurs when a stimulus that was previously neutral to the animal takes on the power to elicit one of these reflexlike responses. In operant conditioning we are looking at observable movements that the animal makes by virtue of skeletal muscle actions. The learned behavior may be anything that the animal is physically capable of performing. In classical conditioning the responses that can be conditioned to a neutral stimulus are limited to those that have already been preprogrammed into an animal's nervous system. We shall first deal with classical conditioning.

Most people are familiar with the paradigm for classical conditioning associated with salivation in dogs. This type of learning was made famous through the classical work of Ivan Pavlov around the turn of the century when he studied the conditioning of salivation as a model for learning. The term "classical" conditioning actually refers to the work of Pavlov as being classical in the field.

Classical conditioning involves the occurrence of a neutral stimulus immediately prior to the unconditioned stimulus that evokes a visceral response. After one or more pairings, depending on the type of response being conditioned, the neutral stimulus will eventually evoke the visceral response alone. The paradigm for this type of learning is given below:

Unconditioned Stimulus (food) – – – – – → Response (salivation)

Conditioned Stimulus (bell) – – – Unconditioned stimulus (food)
– – – – → Response (salivation)

Conditioned Stimulus (bell) – – – – – – – – → Response (salivation)

Conditioning is most efficient when the neutral stimulus precedes the unconditioned stimulus by an interval of only a few seconds or less than a second. This type of conditioning involves responses normally under an animal's "involuntary" control. For example, without the use of a meat powder or some other suitable food, it would be virtually impossible to train a dog to salivate on verbal command.

Pavlov's classical example of conditioning a dog with a bell to evoke salivation is a useful example for our purposes because it introduces the notion of conditioning an internal state in an animal to a previously neutral stimulus. We can assume that when a dog is salivating for food, the internal emotional state that we cannot observe is pleasurable or attractive to the animal. The physiological parameters involve an enhancement of activity in the parasympathetic nervous system and some feedback from activity in the viscera which the animal finds rewarding. We will refer to this internal state as an appetitive emotional reaction. One would suspect that in dogs it is often accompanied by tail wagging and facial expressions that owners recognize as happy. Thus in Pavlov's classical example of a bell acquiring the ability to evoke salivation, we can assume that the bell also evokes an appetitive emotional reaction that is rewarding to the animal.

A variety of visceral responses can be classically conditioned. These include milk ejection, which is commonly conditioned in cattle to the banging of milk cans; secretion of insulin, which is normally elicited by ingestion of sugar but can be conditioned to the smell of food; nausea, which is elicited by food poisoning but can be conditioned to the taste or smell of the food associated with the poisoning; and asthmatic reactions, which are normally caused by foreign proteins, but can be conditioned to the environment in which the foreign protein is administered. Responses of particular interest are the fear and anxiety that are normally evoked by painful stimulation but can be conditioned to the sight of a person or an environment associated with administration of the pain. Painful stimulation in an animal usually causes an increase in blood pressure, sweating, slowing of intestinal motility, and a variety of other visceral reactions which stem from the release of hormones from the adrenal medulla (epinephrine) and the adrenal cortex (corticosteroids). On the outside the animal shows responses we usually associate with fear or anxiety; we can assume that the internal emotional state felt by the animal is aversive or unpleasant. We will refer to this as an aversive emotional reaction.

Although pain will innately evoke an aversive emotional reaction, some other stimuli will also. These include some types of restraint, a sudden change in stimulus intensity such as falling, and intense physical stimuli such as loud noises. Slapping a puppy with a folded newspaper probably produces an aversive emotional reaction because it produces an intense auditory stimulus, not because it is painful. Neutral stimuli that accompany or slightly precede a stimulus that innately evokes an aversive emotional reaction may be conditioned by only one or two pairings. A dog that is in a single automobile accident may, thereafter, refuse to go in the automobile. The automobile, which was previously a neutral stimulus, was conditioned to the aversive emotional reaction. In one case we know of, this presented a severe problem for a blind owner who heavily depended on her dog and had taken the dog to work everyday in an automobile.

It should be noted that both habituation and classical conditioning play a role in shaping fear and anxiety reactions. Some stimuli, such as loud sounds or strange objects, would have some likelihood of being associated with potentially harmful situations in the natural environment so these stimuli innately evoke fear responses; but since they are not painful in themselves, they may be habituated. Neutral stimuli that do not have much of a likelihood of being associated with painful or harmful stimuli in nature may be classically conditioned to fear responses and the accompanying aversive emotional reactions. Thus, habituation attenuates emotional responses, and classical conditioning extends the range of stimuli that evoke aversive emotional reactions.

Because of the unpleasant aspects of aversive emotional reactions, an animal will try to attenuate the reactions by driving away the conditioned stimulus that evokes the reaction. Hence, we may find threatening or fear biting behavior in dogs if the behavior has paid off by driving away the fear inducing stimulus.

Fear reactions may be conditioned to the handling of parts of an animal's body. An inflammatory process that involves, for example, a rectal fistula, is naturally painful for an animal. While the inflammation is present, the animal may respond with aggression when it is touched or handled in the rear quarters. After the inflammatory process has been eliminated by medical treatment, the conditioned response to handling the rear quarters may remain, and the animal will react with excitement and aggression.

Normally, extinction of classically conditioned responses occurs when the neutral stimulus is presented over and over again without being followed by the stimulus that innately evokes the response. This is easily seen regarding an appetitive emotional reaction evoked by a neutral stimulus associated with food. If the conditioned stimulus is no longer followed by food, the connection is extinguished and the conditioned stimulus no longer evokes the appetitive emotional reaction.

With regard to extinction of classically conditioned fear responses, the picture is not so clear. A conditioned stimulus that evokes a fear reaction also evokes attempts on the part of the animal to escape from the conditioned stimulus and the aversive emotional reaction that it creates. These escape attempts, if they include breaking through glass doors, or chewing through fencing, are accompanied by pain. Thus, a neutral stimulus actually leads to pain. Also the internal feelings of aversive emotional reactions are unpleasant. Thus, it is difficult to extinguish the conditioned fear reaction when a full-blown fear response is evoked. However, the process of presenting the conditioned fear-inducing stimulus at very low intensities, which evokes only a mild fear reaction, can lead to extinction. Once the fear response is extinguished at the mild level, the intensity may be increased. This process is continued until the conditioned stimulus may be presented at full strength. The process of presenting stimuli in gradually increasing intensities is called systematic desensitization, the same process that is used for habituating innate fear reactions. Sometimes it is not feasible to diagnose whether fear reactions are actually unhabituated responses or classically conditioned responses. Fortunately, the treatment of systematic desensitization is the same for each.

Operant Conditioning

The concept of operant conditioning is quite simple. If a response is followed by a reward, usually called an reinforcer, the probability of the response occurring again increases. For example, when begging behavior occurs at the table and is followed by a reinforcement such as a piece of cheese, begging behavior tends to be repeated. Operant conditioning is responsible for a number of obnoxious behaviors in dogs, such as jumping up on people, chasing cars, digging holes in cool flower beds, and engaging in a variety of attention-getting behaviors. Operant conditioning may enhance the objectionable nature of fear reactions shown by dogs to gunshots and thunderstorms and it may intensify fear biting. Operant conditioning is also used to train animals in a number of useful and desirable behavioral patterns. Most circus performing animals are trained on the basis of operant conditioning. The paradigm used for representing operant conditioning is presented below.

Stimulus · · · · · Response – – – – → Reinforcement

The paradigm points out that when a response is made in a certain situation (stimulus) and it is rewarded (reinforcement), the response tends to be repeated or maintained with increasing frequency. Operant conditioning concerns voluntary responses involving skeletal muscles. In this respect it differs from classical conditioning, which involves the smooth muscles and glands that are normally associated with involuntary responses.

Positive and Negative Reinforcement. Reinforcements for operant conditioning may be positive or negative. We usually think of rewards or reinforcement as positive. Reinforcers such as food, water, sexual activity, and exploratory behavior are positive, which means that their presentation will result in learning. With most positive reinforcers, such as food, it is required that an animal be deprived of it before it can be reinforced. Sleep is reinforcing only to a sleep-deprived animal. Finding an appropriate place to eliminate is reinforcing to an animal with a full bladder or rectum. Some positive reinforcers clearly do not require deprivation. This may be true of highly favored foods, such as hamburger or cheese, which may be reinforcing to dogs after they have already consumed a large meal. Social contact and petting are also examples of stimuli that are generally reinforcing for dogs and some cats and require little or no deprivation.

Negative reinforcers are aversive stimuli which, when removed, increase the probability of the response. Termination of pain or the reduction of fear are negative reinforcers. Animals learn tasks including escaping or acting aggressive if such behavior reduces the pain or fear which began prior to the response. One can see the process of negative reinforcement at work when we observe an animal that is approached by someone who evokes fear. The animal, by threatening or snapping, drives away the person, which leads to a reduction of the fear and the accompanying unpleasant emotional reaction. Hence, the behavior tends to be repeated.

Negative reinforcement is easily confused with punishment. Punishment is the presentation of an aversive stimulus such as pain, intense noise, or social isolation after an undesirable behavior has occurred.

When punishment is delivered after a behavioral act, it tends to stop the ongoing act.

Operant conditioning techniques work well to enhance a behavioral pattern that is already in an animal's repertoire. Scratching at a door occurs every once in a while without any conditioning. This is what we call the operant level of the response. When the door scratching is reinforced by allowing the dog in, the probability of door scratching increases.

Successive Approximation. Operant conditioning techniques can also be used in behavioral therapy. It is perhaps best to consider examples of teaching parlor tricks that are not normally in an animal's repertoire, such as rolling over or jumping through a hoop. The technique is that of successive approximation. In the use of successive approximation, a behavioral goal is derived and a starting position is determined. One can then define a gradient from the starting point to the goal. The animal learns operants that are successively closer to the goal. In teaching a cat to jump through a hoop, for example, one would initially place the hoop on the floor and stand on the side of the hoop opposite to the cat, say jump, and when the cat walks through the hoop give it a food treat. This is repeated until the cat readily walks through the hoop. We now have a starting point. The process of successive approximation is then used when the hoop is raised a few inches from the floor and the cat is asked to jump through the hoop again and given a food treat every time it walks over the raised hoop. After a few trials at this level, the hoop is raised again a few more inches and the cat is again rewarded every time it goes through the hoop. Over a series of many trials the hoop is very gradually raised until it can be a foot or two off the floor.

The process of successive approximation was described earlier in shaping a cat to be properly toilet trained (see Fig. 13–1). This simple procedure takes advantage of the cat's natural tendency to urinate and defecate in a sandy area. If all goes as planned, one has the ultimate solution in training cats to be properly toilet trained.

The key to the use of successive approximation is to determine the beginning point and define a goal as specifically as possible. One must devise a gradient along which the criterion for rewarding an animal's performance is gradually raised. In using successive approximation to shape behavior, the animal should be rewarded with praise, affection, and favored food treats, each time it behaves appropriately.

Extinction and Reinforcement Schedules. Once a response has been learned, it is generally maintained as long as reinforcement is at least occasionally presented. If the reinforcement is withheld permanently, an animal will make fewer and fewer responses until the behavior drops to the previous operant level. This process is called extinction. Note that extinction is an active process, requiring behavioral responses to be made. Extinction should be distinguished from forgetting, which is a passive process representing the elimination or decrement of response strength through the passage of time. A feigned lameness, which is maintained because it evokes attention from the dog's owner, would eventually be extinguished if the owner decided to withhold all attention from the dog whenever the behavior was exhibited. Extinction, of course, would work with many of the obnoxious behavioral patterns that dogs show, such as scratching on screen doors and begging for food, if the behavior is never again reinforced.

Sometimes extinction is impossible because the reinforcing aspects of the behavior cannot be removed from the behavioral act. For example, when dogs jump up on people the reinforcement is built into the response since by jumping the dog gets close to people and makes contact with them. Another behavior that has a built-in reinforcement is digging holes in the backyard or being destructive in the house if the rewarding payoff involves relief or escape from boredom. These behavior patterns are solvable by other means, namely punishment and counter conditioning.

Success in treating objectionable acquired behavioral patterns through the process of extinction requires an understanding of the relationship between the reinforcement and maintenance of the behavior. This leads to a discussion of intermittent versus continuous reinforcement. In teaching an animal a simple parlor trick, one usually attempts to reward with food or praise each response or short series of responses. This one-to-one reinforcement schedule is often referred to as a continuous reinforcement schedule. After an animal learns something, it is not necessary that it be rewarded each time; intermittent reward can usually maintain the behavior. Consider the behavior of a dog scratching a door to be let in. The animal is obviously putting out many scratches, perhaps 100, before it is allowed in. If one wanted to extinguish door scratching, the behavior could be eliminated by withholding permanent reinforcement. Undoubtedly, thousands of responses would be made by the dog before its behavior would be extinguished.

Behavior that has been maintained on intermittent reinforcement is much more resistant to extinction than that maintained on continuous reinforcement. If one wanted to extinguish door scratching in a dog, the extinction would be easier for a dog that had been rewarded almost every time it started door scratching (by being allowed in the house) than it would be for a dog that was rewarded only after the scratching became very frequent and intense. What usually happens, however, is that milder forms of door scratching have been extinguished because they never paid off and the more intense and frequent forms of scratching were gradually shaped because this was the behavior that yielded reinforcement. Thus, shaping through successive approximation often occurs when behavior is reinforced on an intermittent schedule.

There are two types of intermittent reward schedules. One of these is a ratio. If an animal must make an exact number of responses before it is reinforced one time, then it is on a fixed ratio schedule. If the number of responses per reinforcement is variable, but presumably on an average, the animal is being reinforced on a variable ratio schedule. In practice most reinforcement is given on a variable ratio rather than a fixed ratio schedule. On both fixed and variable ratio schedules, operant learning behavior is emitted rapidly since the sooner the required number of responses are made, the sooner reinforcement is attained.

The other type of intermittent reinforcement is the interval schedule. This may also be fixed or variable. In a fixed interval schedule an animal is not reinforced until a specific amount of time has elapsed since the preceding reinforcement. We commonly see this in animals that show waiting behavior. The dog that goes to the door at 8 o'clock each evening slightly before it is taken out on its routine 8 o'clock walk is being reinforced on a fixed interval reinforcement schedule. Animals have excellent biological clocks, and their behavior is maintained well on a fixed interval schedule. If an animal is reinforced on a variable interval schedule, it means that the duration of time elapsed since the preceding reinforcement varies but may be represented by an average interval duration. Animals that are reinforced on this type of schedule are likely to display waiting behavior on a continuous basis. We use this reinforcement technique in conditioning dogs to stay around the yard when a food treat is given at variable times throughout the day.

THERAPEUTIC TECHNIQUES

Therapeutic techniques are derived from conditioning processes and are designed for specific problem behaviors. These techniques include systematic desensitization, counter conditioning, extinction, and punishment.

Systematic Desensitization

Systematic desensitization is usually applied to phobias or fear reactions and allows the habituation of innate phobias or the extinction of classically conditioned responses to occur gradually. Physiological reactions that generally accompany fear and anxiety are in themselves aversive so that stimuli that evoke full blown fear and anxiety reactions are difficult to habituate or extinguish even if the stimuli cause no actual pain. However, if stimuli are presented gradually, along a gradient of distance or intensity, the stimuli eventually lose their ability to produce an emotional reaction (desensitization). What is actually happening is that the habituation or extinction occurs to a low level stimulus and this is generalized to stimulus intensities slightly higher than the stimulus presented. The stimulus intensity may then be increased slightly without evoking a noticeable emotional response. The desensitization is then carried out at that new level and this generalizes somewhat, allowing one to raise the stimulus intensity again. If this procedure is continued until the stimulus is presented at full strength, the fear reaction is eventually completely habituated or extinguished to that stimulus.

This technique is referred to as desensitization because the intense emotional response to the stimulus is desensitized, much as one might desensitize an animal to an allergen or antigen. The term systematic is used because the stimulus is presented at gradually increasing degrees of strength on a very systematic and orderly basis.

Systematic desensitization is useful in treating the various fear reactions or phobias which include natural or acquired fears of sounds or strange people, fear biting, submissive urination, and separation anxiety.

It is important to be aware of the types of gradients that are used in systematic desensitization. When designing a program of training sessions, it is critical to build in a gradient with increments that gradually increase the stimulus intensity for the animal in some measurable or logical fashion. At the beginning of the gradient, the animal should experience a low level, tolerable stimulus, which if presented in full strength, would be sufficient to evoke the emotional reaction. In deciding on the stimulus to be used, one must first present the stimulus close to full intensity to make certain that the stimulus does in fact evoke the undesirable emotional reaction. For example, in using a commercial recording of a thunderstorm to desensitize a dog to thunderstorms, one must play the thunderstorm record at full strength to make certain the stimulus is adequate to evoke the response. The same thing goes with using muffled starter pistol sounds to desensitize a dog to gunshots; the unmuffled starter pistol must be fired out at least once to make sure that it will evoke the same response as loud gunshots. After performing this first critical test, one must find a way of reducing the stimulus intensity along a systematic gradient.

The first published account of the use of systematic desensitization to treat problem behaviors in dogs was by Tuber, *et al.* (1974) who applied this technique to an Old English Sheepdog that had acquired an intense fear of thunderstorms. The dog's behavior was dangerous to itself and its owners. At the first indication of an impending storm, the dog would begin accelerating its aimless pacing, exhibit profuse salivation, and marked panting. These responses were climaxed by the dog throwing his body against any obstacle in an attempt to escape.

Systematic desensitization of fear reactions to certain people such as strange men can also be approached on a graded basis. Training sessions can be planned in which a man who normally evokes an emotional response enters a room at a distance far enough from the animal so that the man is seen but the emotional reaction is not evoked. The man makes an appearance at a given distance, approximately 10 times in a row, and this constitutes a training session. Over subsequent sessions the man moves slightly closer each time. The gradient being used here is distance.

An account of the use of this systematic desensitization to deal with fear of people can be found by Voith (1980b).

On occasion it is impossible to attenuate a stimulus sufficiently to produce only a mild or no emotional response. Therefore, one does not have a starting point on a gradient. Blocking an emotional reaction by chronic administration of a tranquilizer may allow one to gain a starting point on a gradient so the stimulus intensity may be increased gradually over several sessions. The tranquilizer can then be gradually phased out and systematic desensitization continued. The use of a tranquilizer as a adjuvant in this type of therapeutic conditioning is discussed in Chapter 24.

Systematic desensitization is almost always paired with a counter conditioning technique to accomplish the most effective therapy for phobias. The use of counter conditioning as described simply potentiates the desensitization process, but these two techniques are discussed separately because they each utilize different conditioning principles.

Counter Conditioning

Although this technique is used along with systematic desensitization, the emphasis is on establishing a new response, one that is basically incompatible with the undesirable behavior.

Emotional Reactions. Unhabituated fears and aversive emotional reactions acquired through classical conditioning may be reduced when a new appetitive emotional response is classically conditioned to the same stimulus that evokes the fear or anxiety. The appetitive emotional reaction is usually conditioned to a mild form of the same stimulus that evokes the fear or anxiety by pairing the stimulus with food treats and affection. Thus, the process of classical conditioning can be used to setup an appetitive emotional reaction that will eventually come to replace the aversive emotional reaction, providing that the stimulus that evokes the aversive emotional reaction is desensitized.

Both counter conditioning and systematic desensitization are conducted simultaneously. As an illustration, take the example of a dog that is very fearful of the sound of gunshots. This fear may be an unhabituated emotional reaction or could be a classically conditioned response if the dog had been, in fact, shot with a gun at one time. Regardless of the actual diagnosis, the therapeutic approach is the same. We can arrange training sessions of 10 trials to expose the dog to the sound of gunshots that are muffled with several layers of cardboard boxes. One might use a starter pistol that fires .22 caliber blank cartridges and muffle the gunshots by a series of nested cardboard boxes as shown in Figure 4–1, page 63. If the gunshot is sufficiently muffled, the dog can be called over to sit near the box and when a shot is fired, its emotional disturbance will be mild. We then start the counter conditioning by giving the dog a bit of favored food after each shot of the starter pistol. It is common practice for a training session to consist of 10 gunshots. The food will create an internal appetitive emotional reaction that is classically conditioned to the muffled gunshot since food follows the stimulus, and after a couple of training sessions the stimulus itself will come to produce an appetitive emotional reaction. This reaction is incompatible with the aversive emotional reaction associated with the fear response. In our training sessions the fear response is evoked only mildly, if at all, because the stimulus is too weak. As the nested cardboard boxes are removed, the muffled gunshot becomes louder but the degree of aversive emotional reaction produced is weak. With each session the animal's emotional response to the gunshot is desensitized while the appetitive emotional reaction is continuously conditioned to the stimulus. At each new level in which the muffling is removed, desensitization continues and counter conditioning is maintained. Eventually the gun can be fired close to the dog with no muffling and instead of evoking the fear reaction as in the past, an appetitive emotional reaction is produced.

Although it is common for us to think of counter conditioning as useful in phobias and anxiety reactions, the process is useful for other types of behavioral therapy as well. The establishment of an appetitive emotional reaction through counter conditioning may be used to create a favorable disposition in a pet to people that it is neutral toward or does not like. We can use another case to illustrate this. A dog had been presented for consultation because it frequently threatened and occasionally had bitten the wife of a couple. The dog had been strongly attached to the man who had the dog before the marriage. His wife accepted and liked the dog, but the dog had an obvious dislike for her and challenged her authority from time to time. As is usually the case, the dog was punished by the husband for this aggressiveness but the results were not lasting. Counter conditioning was employed when the husband was instructed to withdraw all of his affection and attention from the dog for a two-week period. The dog could obtain only praise, affection, and favored food treats from the woman. Often in such instances a 24-hour food deprivation is useful. This placed the woman in a much more favorable position. Since it was she who administered the rewards, she acquired the advantage of evoking the appetitive emotional response normally associated with these rewards. When the dog approached the man for some attention, he simply turned away so that the dog would approach the woman. The woman was instructed to take advantage of these opportunities to gain the upper hand by requiring the dog to respond to some commands, such as "sit" or "lay down," before giving the dog reinforcement.

Counter conditioning has also been used to deal with separation anxiety. Because of the fear and anxiety produced by being left alone, a dog engages in a variety of misbehaviors, including defecating and urinating in unacceptable places, vocalizing excessively, and being destructive. One therapeutic approach consists of giving the dog a large bone as the owner leaves for short periods of time. The dog will associate being left for brief periods with the reinforcing properties of chewing on a bone. The bone creates an appetitive emotional reaction that is strong enough to overcome the mild anxiety that follows a short departure. If the dog is left for a very long period of time, the anxiety may build up sufficiently to overshadow the emotional reaction associated with the bone. Therefore, in the initial stages of training, the dog can be left alone only for short periods of time. As training sessions progress, the dog is left alone for longer and longer periods. It is gradually desensitized to the stimulus of being isolated.

Operant Responses. Counter conditioning approaches can also involve the operant conditioning of a response that is incompatible with an undesirable operant response. If a dog is digging holes in the backyard and the digging behavior does not stem from separation anxiety but from something like boredom, then one might solve a digging problem by training the dog to dig in a spot that is acceptable. If a dog is running away from its home on a daily basis, the problem might be solved by training a dog to stay around the house for food treats given intermittently throughout the day. Staying around is obviously incompatible with running away.

The use of operant conditioning techniques in counter conditioning usually involves shaping a desirable response through the process of successive approximation. Thus, it is usually necessary to define the behavioral goal in precise terms, determine a beginning point for conditioning, and devise a gradient from the beginning point to the behavioral goal. Teaching a dog to stay around the backyard to solve a roaming problem illustrates this principle. Initially, one could go out into the backyard every 5 or 10 minutes and reward a dog for being there with a food treat. The behavior being rewarded is simply hanging around and may take different forms such as sleeping, playing, exploring, and watching automobile traffic. The average time between rewards is gradually increased and the intervals are varied (variable interval schedule).

A final example in the use of operant conditioning in counter conditioning is teaching a cat to ring a bell to get the owners to open the door to allow it to come in. This incidentally is a cure for the problem of scratching at a screen door. A bell, mounted on a bracket, can simply be placed close to the spot where the animal is scratching so that it cannot help but hit the bell on occasion. The owner simply waits inside for the animal to hit the bell but as soon as that occurs, the cat is immediately let in. All other scratching behavior is ignored. After a few days most cats (and dogs) will be hitting the bell rather than scratching at the door. Over a period of the next few days the bell may be gradually moved over to the side of the door to a final desirable location. Note that successive approximation must be used to shape bell-ringing behavior.

Since bell ringing is the only response that produces door opening, the bell ringing is incompatible with door scratching and the problem is solved.

Extinction of an Operant Response

Under the heading of systematic desensitization, we dealt with the elimination of undesirable emotional responses. The processes were either habituation or extinction of a classically conditioned response. Under this heading, we are concerned with the extinction of an operant response that does not require the gradual approach used in systematic desensitization. One of the techniques to eliminate an undesirable operant response is extinguishing the behavior by permanently withholding any further reinforcement. If this is done the behavior will eventually be reduced to a frequency no greater than the operant level. Extinction is, of course, easier with responses that have been reinforced on a continuous rather than an intermittent basis.

One of the best uses of the extinction technique is in dealing with attention-getting behaviors such as feigned lameness, snapping at imaginary flies, or tail chasing. By simply walking away whenever the animal shows the attention-getting behavior, the behavior is eventually extinguished. One can also use some counter conditioning by rewarding behavior such as lying still, which is incompatible with attention-getting acts. An example serves to illustrate the point. A client had complained that his Siamese cat was biting on its tail as though it were irritated and itching. No dermatologic problem or peripheral neuritis could be found. Nevertheless, the cat continued to chew on its tail, which became so mutilated that it had to be amputated where the cat could no longer reach it. Still the cat attempted to grab at its phantom tail. Careful questioning of the owner revealed that the cat only chased its tail in the owner's presence, and when it did so, the owner invariably picked up and comforted the cat. The client was instructed to simply walk away from the cat whenever the behavior occurred. Several days later the client reported that the recommendations were effective. For the first few days, when the client walked away from the cat, the animal would sometimes follow him into the next room and begin his little act again. The owner simply walked away a second time. Over about two weeks the tail chasing behavior disappeared.

Extinction is almost impossible to use for cases in which a reward is closely linked to the behavior and cannot be separated. This would be true for a dog jumping up on people. The rewarding aspect is built-in, because as soon as the dog jumps up, it receives interaction with a person. For instances in which extinction is not feasible, the best approach with undesirable operants is to counter-condition alternative behaviors that are acceptable as discussed above.

One of the interesting aspects of extinction is that at the initial part of the extinction process, an animal will engage in the behavior more frequently and intensely before it tapers off, as if it were frustrated. When behavior no longer pays off, an animal has a tendency to try harder and faster for a period of time before giving up. Therefore, in giving clients advice and in using extinction, it must be emphasized that for a period of time after the reinforcement is withheld, the behavior may get somewhat worse before it gets better.

Another point to emphasize when prescribing the extinction technique is that once a pet owner decides to start extinguishing a behavior, the withdrawal of the reinforcement must be complete. One must not give in, even occasionally, to relieve the temporary objectionable behavior. If one gives in, the behavior then becomes reinforced on an intermittent schedule and at the same time the more objectionable parts of the behavior are shaped. By giving in even once or twice, one makes the whole extinction process much more lengthy and unpleasant. Extinction is basically an active process and one is simply waiting for a number of responses to be emitted before elimination of the behavior finally takes place.

PUNISHMENT AND AVERSION CONDITIONING

Punishment is a complex process that includes using elements of an animal's innate behavioral predispositions, classical conditioning, and extinction. In its purest form, punishment simply stops ongoing behavior and does not extinguish or counter-condition a response. We use punishment with the intention that the effects will endure, that the animal will learn something from the experience. When this actually happens there is usually some sort of conditioning process involved. Punishment means different things to different people. For some it simply means hitting an animal for doing something wrong, for others it may mean using an aversive stimulus that is not associated with the person administering it (remote punishment). If we walk away from a dog every time it is engaging in an attention-getting act, we may think to ourselves that we are punishing it by walking away whereas, in fact, we are extinguishing an operant response. In this section we will discuss three techniques that are used to stop behavior that is undesirable. These are interactive punishment, remote punishment, and aversion conditioning. All of these techniques are referred to from time to time as punishment. It is important to distinguish between the techniques because the particular technique that we would recommend has to be indicated for the problem we are trying to solve.

Interactive Punishment

In interactive punishment a person hits an animal with his hand or an instrument such as a rolled up newspaper, shouts at it, or throws something at it in a way in which it is obvious that the aversive stimulus is coming from the person. Thus, the animal clearly associates the unpleasant stimulus with a person.

Interactive punishment is clearly indicated when owners must assert their dominance over their dogs to maintain acceptable dominant-subordinate relationships. Growling or snapping at the owner, when it is not a reflection of fear, is best met with force. Dogs are social animals that respond naturally to factors that maintain a dominance hierarchy. Insufficient dominance is one of the most common behavioral problems of dogs and their owners, and it often relates to lack of assertiveness on the owners' part.

Aside from dealing with aggression, interactive punishment is commonly used in pets to correct misbehaviors that occur in the owner's presence. This is usually effective to the extent that a dog will learn not to engage in the behavior when the owner is present. With both dogs and cats, it is probably fruitless to administer interactive punishment hours or even several minutes after an act of misbehavior. There is only a slight chance that the connection will be made between the punishment and the punished act. One of the other common difficulties in the use of interactive punishment is inconsistency on the part of the owner. Jumping up on people by dogs, for example, may be allowed or encouraged on the weekend when the owner has on old clothes, but punished on weekdays when the owner is dressed for work. Because the punishment and reward are administered on an intermittent basis the behavior is maintained because the behavior is still rewarded.

Remote Punishment

The most effective type of punishment for acts of misbehavior is remote punishment. One of the best examples is the use of the remote-controlled shock collar. A battery-powered capacitor delivers an electric shock through electrodes that are mounted on the collar. The capacitor is discharged through a remote control unit that is held by the owner. When the shock collar is used properly, the aversive stimulation is associated with the act of misbehavior, and with the object to which the behavior is directed, because a shock can be delivered immediately following it. Shock collars have proven effective in a variety of behavioral problems in dogs, including chasing cars and bicycles, and running away when being called. A disadvantage of the shock collar is that a "dummy" collar must be used periodically during training procedures so that the animal does not make the association of the collar weight with the shock. Experience has shown that if the shock collar is not effective in altering behavior in the first few trials, it probably will not cure the misbehavior with repeated trials. While most practitioners would not find electric shock treatment appropriate to use in the time available, Tortora (1983a; 1983b) takes dangerously aggressive dogs under his full time care and incorporates electric shock within an intensive rehabilitation program. Others also find electrical shock useful (Vollmer, 1979a; 1979b; 1980).

Another type of remote punishment that is effective for specific acts is the use of mousetraps with dogs large enough not to be injured by them. Instead of hitting or scolding dogs that jump on household furniture or dig in a flower bed, one can place mousetraps in the areas where the dog is likely to approach (Fig. 22–1). The loud snap of the trap and the sting, if the dog gets caught, are aversive and the behavior is punished immediately and by the object to which the behavior is directed.

Mousetraps that are spring-loaded and then carefully set upside-down can be used with cats (Fig. 22–2; Hart, 1979b). When a mousetrap is touched and then flies up into the air, the loud snap is usually enough to frighten a cat. Mousetrap therapy has been used to remotely punish feline problems such as digging into flower pots, jumping on counters, soiling a child's sandbox, digging into flower beds, and urine spraying.

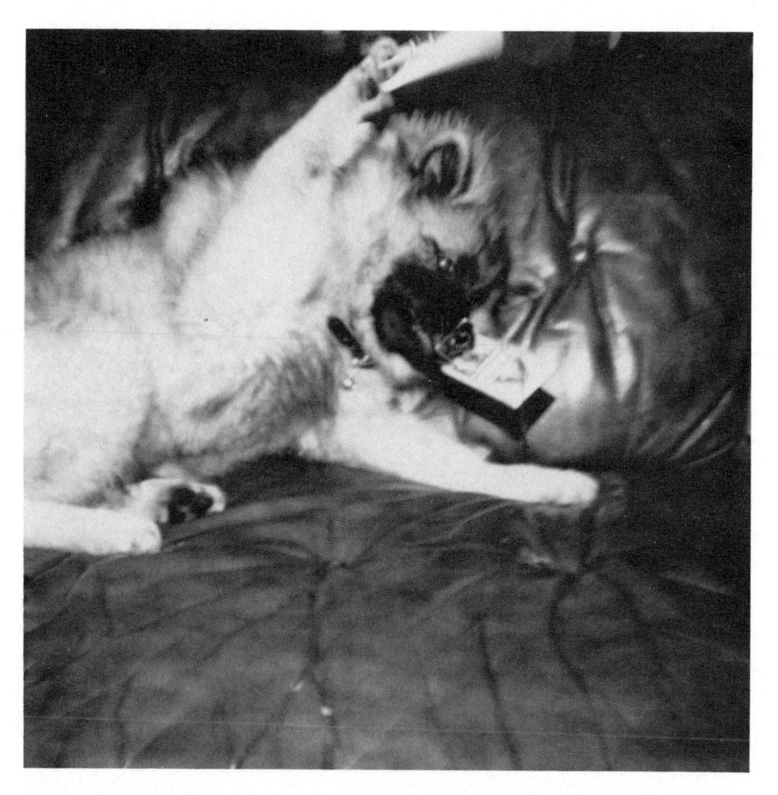

Fig. 22–1. Loaded mousetraps are a form of remote punishment for large or medium size dogs. Punishment is delivered immediately and by the object to which the misbehavior is directed—in this case the couch.

Fig. 22–2. Remote punishment for cats and small dogs can be achieved with mousetraps carefully placed upside down where they are triggered by slight movement. (From Hart, 1979; reprinted with permission, Veterinary Practice Publishing Co.)

Most mousetraps must be adjusted so that they can be set and placed upside-down and not go off until touched. This usually requires some adjustment of the tongue that holds the cheese. One can use single mousetraps for small places or arrange a group of them in a line with the intention of achieving a chain reaction. The most interesting arrangement is to stack three or four mousetraps together (Fig. 22–3) so that they all go off at the same time. Mousetraps can also be used in the hanging position. One might be able to use a ring of mousetraps around the trunk of a tree to deal with a cat that consistently climbs up a certain tree and has to be rescued repeatedly.

As with any type of punishment, it is important that the use of remote punishment be very consistent. As soon as possible after a set of mousetraps has been set off, they should be reset so if the misbehavior is repeated it is also punished. After a few days, if there is no evidence of further misbehavior, it may be possible to remove the mousetraps.

The use of water sprayers is another type of remote punishment, especially good for cats. But certain specific instructions must be followed. The common plant sprayer can be adjusted so that it delivers a mist or a stream of water that is usually more powerful than waterguns. For remote punishment one uses the solid stream. To be effective, the cat must be sprayed with the water with complete detachment. The animal must not see the person spraying it. One, of course, must always be around to do the squirting of the cat and this is one of the disadvantages of water sprayer therapy when the animal is engaging in destructive

Fig. 22–3. In this example of remote punishment to discourage use of a plant as a toilet area, one loaded mousetrap is stacked on top of two other loaded mousetraps, which in turn are placed on a piece of cardboard that supports the traps for springing when triggered. The mousetraps are somewhat camouflaged by sprinkling dirt on top of them. (From Hart, 1979b; reprinted with permission, Veterinary Practice Publishing Co.)

behavior throughout the house. This may happen, for example, if the animal is eating various plants throughout the house. It may be necessary to physically isolate the cat from areas of the house in which it does damage until the owners are around and able to punish the misbehavior. When the owner notices the cat engaging in misbehavior, he should approach the cat slowly in an unemotional fashion, grab a water sprayer that had been placed around the house for the purpose and, without being identified as the source of the punishment, let the cat have it. When sprayers are left at several locations throughout the house, it will not be necessary for the owner to dash around looking for the sprayer hoping that the cat will stay at the misbehavior until the sprayer is found.

Aversion Conditioning

We all recognize that punishment clearly stops ongoing, undesirable behavior. This is a primary motivation for using punishment. In addition, punishment almost always involves aversion conditioning. In fact, this is the reason punishment of a behavioral act carries over into the future. Punishing a dog with mousetraps for jumping on a couch, for example, produces an aversion to the couch because proximity to the couch is paired with the aversive properties of the snapping mousetraps. Proximity to the couch, which was previously a neutral stimulus, now evokes an aversive emotional reaction by virtue of classical conditioning. Thus, punishment has the effect of stopping ongoing, undesirable behavior and producing an aversion to the object associated with the misbehavior. This is the reason, incidentally, that interactive punishment is not recommended for simple acts of misbehavior; the ongoing undesirable behavior is stopped, but an aversion may be produced to the person administering the punishment rather than the object toward which the misbehavior was directed.

Since aversion conditioning is an effective technique in eliminating some types of problem behavior, it is important to realize that it need not be part of the punishment. One does not need to wait for misbehavior to occur to use aversion conditioning. If a dog is killing chickens, cats, or sheep, aversion conditioning to these prey can be instituted without waiting for another animal to be killed. The conditioning might consist of placing a sheep leg with skin attached (which was previously frozen for the occasion) in the dog's mouth and taping the mouth shut for one hour per day. Within a few days this process could produce an aversion to live sheep, especially if the lamb leg is amply laced with cayenne pepper. However, one does not need to wait for a lamb to be killed by the dog before running each aversion conditioning trial. Rather, one can systematically go about producing the aversion and the process is much more efficient than conducting an aversion conditioning trial only after each misbehavior.

The main difference between punishment and aversion conditioning is that in punishment the aversive stimulus is linked temporally to a misbehavior, but in aversion conditioning, an aversive stimulus is used systematically on a scheduled basis. The old trick of hanging a chicken around a dog's neck when it killed a chicken was thought of as punishment. However, the idea was to leave the dead chicken on the dog until the chicken rotted, figuring that the dog would get "sick" of it. In reality, a form of aversion conditioning was being used, and any lasting effects of this treatment were undoubtedly due to aversion conditioning. Once this is clear, it becomes obvious that the appropriate therapy is not so much to punish the misbehavior, but to carry out a scheduled aversion conditioning program as efficiently as possible.

TABLE 22–1. Classification of Conditioning Procedures

Process	Technique	Goal	Indications & Applications
Innate Emotional Response			
Habituation	Systematic Desensitization	Eliminate innate aversive emotional reaction	Submission urination; innate phobias; fear biting; fear of strangers
Classical Conditioning			
Extinction	Systematic Desensitization	Eliminate acquired aversive emotional reaction	Acquired phobias; acquired food aversions
Acquisition	Counter Conditioning	Establish appetitive emotional reaction	All problems listed above; territorial aggression; competitive aggression
	Aversion Conditioning	Establish aversive emotional reaction	Eating house plants; coprophagy; urine spraying; scratching furniture; killing chickens and cats
	Remote Punishment	Establish aversive emotional reaction	Destructive behavior; barking; urine spraying; scratching furniture; escape from confinement
Operant Conditioning			
Extinction	Extinction	Eliminate acquired operant	Attention-getting behavior; escape and roaming; phobias that are rewarded
Acquisition	Counter Conditioning	Establish operant response	Destructive behavior; dominance-related aggression; roaming

Another form of aversion conditioning can be used with cats that eat house plants, scratch in the wrong places, and eliminate or spray urine in inappropriate areas. The conditioning technique takes advantage of constituents in underarm spray deodorant. Spray deodorant normally contains aluminum chlorohydrate which, when it comes in contact with sensitive mucous membranes, is irritative. Underarm deodorants also contain a perfume that is usually unique to the particular brand. To produce an aversion one can spray underarm deodorant close enough to the nose of a cat or dog that the irritative ingredient gets into the nasal mucous membranes and causes an irritative reaction. This will be associated with the particular smell of the underarm deodorant. After a few trials the animals should develop an aversion to just the smell of the deodorant. One has then only to spray the objects that one wishes to have the animal avoid. Of course an underarm deodorant that is different from that used personally by the owners should be used for aversion conditioning.

CONCLUSIONS

In this chapter we have presented the basic principles of conditioning as they relate to habituation, classical conditioning, and operant conditioning. Behavioral problems may stem from a lack of habituation that normally occurs in the life of an animal or from objectionable behavioral patterns acquired by means of classical or operant conditioning. We use the processes of habituation, classical conditioning, and operant conditioning in the various techniques applied to behavioral problems. Table 22–1 sets forth the various conditioning processes along with the associated therapeutic techniques. A partial list of indications is also given in the table.

Chapter 23

Behavior Modification Through Hormonal Manipulation

The most frequently performed operations in a small animal practice are those which involve removal of the testes from males and ovaries from females. Both testicular and ovarian hormones have pronounced effects on behavior, and these operations comprise the most frequent means of behavior modification in dogs and cats. Male cats are castrated prepubertally to prevent the occurrence of objectionable masculine behavior, including roaming, fighting, and urine spraying. Ovariohysterectomy is routinely performed on females not only to prevent pregnancy, but also to prevent these animals from displaying the behavioral aspects of estrus. Castration of adult male dogs is being employed more frequently in an attempt to alter undesirable masculine behaviors such as urine marking in the house, mounting people, and fighting with other male dogs. This operation is also being employed, especially by animal shelters, in the interest of canine population control. Veterinarians and other professionals are expected to know what behavioral patterns may be altered by this operation.

In this chapter we discuss the endogenous influences of gonadal hormones and focus on therapeutic behavioral modification through hormonal manipulation. It is known from work on laboratory animals that other hormones, such as the adrenal corticotropic hormone, adrenal cortical steroids, and thyroxine, affect behavior. However, these behavioral effects are not pronounced, and there has been little clinical experimentation regarding hormones of other endocrine systems to warrant discussing them. In the future, of course, this may change.

ENDOGENOUS INFLUENCES OF TESTICULAR HORMONES

The behavior of dogs and cats consists of an integrated complex of learned and genetically determined behavioral patterns. Much of our interaction with dogs and cats consists of teaching these animals new tasks and using the principles of learning to alter undesirable acquired behavior. Frequently, it is the innate or genetically determined aspects of behavior that are a source of fascination to pet owners. At other times, inherited behavioral patterns become a source of frustration to dog and cat owners and this brings a practitioner into the picture. Problem behaviors are often those influenced by sex hormones, and the purpose of this chapter is to deal with such behaviors by manipulating the hormonal environment. First, we must take up the topic of sex differences in behavior and the endogenous effects of hormones.

Sexual Dimorphism

Genetically determined behavioral patterns include those common to both sexes, such as play, predatory behavior, greeting responses, tail wagging, grooming, and face washing, as well as behavioral patterns that differ between the two sexes and are, therefore, sexually dimorphic. Females have behavioral patterns related to estrus and maternal care that males do not. Males have behavioral patterns involved in copulation and urine marking that are less frequently seen in females. Male dogs and cats are generally more aggressive than females. Accentuation of aggressive behavior is particularly noticeable in the interaction between males, giving rise to the term, intermale aggression.

The existence of sexual dimorphic differences in behavior does not mean that males and females do not have the ability to display behavioral patterns typical of the opposite sex. The difference between the sexes is one of frequency or probability. Females, for example, can fight as fiercely as males and display mounting and urine marking behavior. At times, males display female sexual behavior. For example, grasping a male cat by the skin of the neck and touching the perineal area sometimes evokes the female receptive posture of pelvic elevation, treading with the back legs, and tail deviation. In Chapter 21 we dealt with some behavioral traits that appear to be sexually dimorphic in a quantitative sense (see Fig. 21–1, page 194). Informants ranked male dogs significantly higher than females in tendency to be dominant over the owner, aggressive to other dogs, and activity level, and female dogs were ranked higher than males in trainability and housebreaking ease.

The behavioral patterns of males that are strongly sexually dimorphic are usually reduced after gonadectomy. However, even after gonadectomy the two sexes remain fundamentally different. The administration of estrogen to castrated males does not generally induce much female sexual behavior, nor does the administration of testosterone to ovariectomized females bring on a full display of male behavior. Females are basically feminine in their behavior and males are basically masculine, as a reflection of relatively permanent differences in the brain.

Although the administration of the hormone of one sex to the opposite sex does markedly change behavior accordingly, there may be some minor side effects. Administration of testosterone, for example, to spayed female dogs, can increase the tendency for females to mount other females, and administration of estrogen to castrated males may make them somewhat sexually attractive to normal males.

Recent research has shown that sex-typical behavioral patterns are basically a function of perinatal gonadal androgen secretion. Fetal and early postnatal androgen secretion in males promotes a masculinization effect in some parts of the male brain, especially the hypothalamus; and the lack of androgen secretion in females leads to the development of the hypothalamus in a feminine direction. Before we consider these developmental effects in more detail, some attention should be given to the pattern of androgen secretion in males.

Testosterone Secretion and Development of Male Behavior

Although it is believed that the chief male sex hormone that influences behavior is testosterone, another androgen, androstenedione, is also involved, but to a lesser extent. Both androgens are produced by the interstitial cells (Leydig's) in the testes. When males have been castrated, the most common androgen replacement used both clinically and experimentally to restore male behavior is some form of testosterone.

Fluctuations in Testosterone Secretions. In adult males of virtually all species studied, including the dog and cat, there are marked fluctuations throughout the day in testosterone. In dogs the fluctuations can be as great as tenfold (Hart and Ladewig, 1980). Usually the peaks in testosterone concentration are preceded by bursts in production of the luteinizing hormone.

The fact that there are fluctuations raises the question of the minimal level of testosterone needed to maintain behavior. Work from cats and laboratory rodents suggests that average testosterone levels are three to four times higher than levels needed to maintain sexual behavior. This finding is consistent with other findings in male mammals which suggests that there is no simple relationship between plasma testosterone and mating activity in male mammals. It will undoubtedly be some time before the behavioral and physiological significance of fluctuations in testosterone are understood. In males of many species, an increase in the secretion of testosterone is seen when a male becomes sexually excited, as when he is introduced to a female. In rats where multimale matings are common, this effect has been related to activation of spinal sexual reflexive mechanisms involved in a male rat's ability to dislodge seminal plugs deposited by other males and to deposit his own tight plug (Hart, 1983). As we learn more about the reproductive biology of dogs and cats, an explanation of the significance of the testosterone fluctuations for these species may be found.

Most mammalian species are seasonal breeders; females come into estrus during a certain part of the year. The adaptive value of such a process is that the young are born at a favorable time. In seasonal breeders there are changes in testosterone in males that correspond to the seasonal cycle of females. This is apparently true to a minor degree in cats. However, in dogs the females are not seasonal in breeding activity and one finds no evidence of seasonality in testosterone secretion in the male.

Individual Differences in Androgen Concentration. The question often arises as to whether the poor sexual activity and breeding performance of some males is due to low testosterone concentration levels in the blood. Therefore, the issue of whether the natural differences in sexual activity one finds in animals are reflected in differences in androgen blood levels is of some clinical interest. Generally it has been found that by giving supra-normal doses of androgen, the degree of sexual activity in laboratory mammals can be exceeded to some extent, but individual differences cannot be overcome (Beach and Fowler, 1959; Damassa, *et al.*, 1977).

Several experiments have shown that there is no relationship between sexual motivation or performance and testosterone levels. The usual procedure is to study animals ranging from low to high in sexual activity, and then castrate the animals and give them all the same amount of replacement testosterone. In subsequent tests for sexual behavior, it has been found that males return to their respective preoperative levels of sexual activity (Grunt and Young, 1952; Larsson, 1966; Harding and Feder, 1976).

When adult male dogs and cats are castrated, all behavioral patterns that are testosterone dependent continue to be displayed for a period of time after testosterone has disappeared from the blood. The persistence of the behaviors for sometimes months, or years, is a reflection of the capability of the brain to continue to mediate this behavior without hormonal support, since following castration testosterone, within hours, almost completely disappears from the bloodstream, as illustrated in Figure 23–1 on cats (Hart, 1979d). The persistence is not due to adrenal gland secretion or other sources of androgen. It has been shown in laboratory animals and dogs that sexual behavior in castrated males persists even after the animals are adrenalectomized (Hart, 1974b).

Development of Behavior and Androgen Secretion

Only recently have investigators been able to measure gonadal hormone levels with some degree of accuracy. In general it is known that at the time of puberty the testes begin secreting testosterone in amounts sufficient to stimulate certain morphologic changes that are recognized as malelike, and to activate male sexual behavior and other masculine behavioral patterns such as urine marking, roaming, and fighting.

In laboratory animals it is known that there is also a transient surge of testosterone secretion in males just before and/or after the time of birth, which is important developmentally. This surge in testosterone

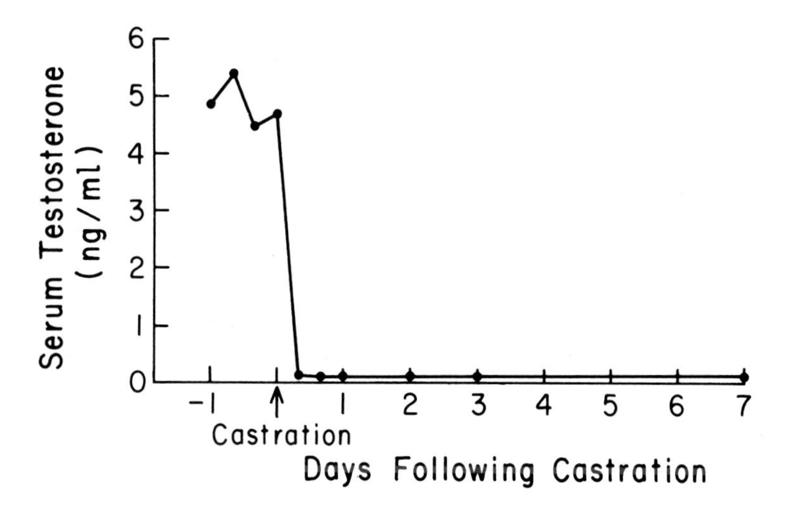

Fig. 23–1. In cats, serum testosterone concentration falls to castrate or nondetectable levels within hours after castration in all animals studied. (From Hart, 1979d; reprinted with permission, Veterinary Learning Systems Co., Inc.)

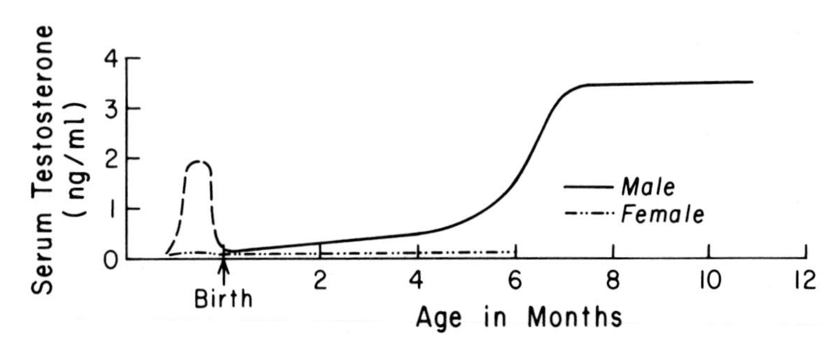

Fig. 23–2. Serum testosterone concentration in male and female dogs from late prenatal stage to adulthood. The prenatal surge for males is hypothetical and based on data from other mammals. The same pattern would be expected in cats. (From Hart, 1979d; reprinted with permission, Veterinary Learning Systems Co., Inc.)

secretion sensitizes certain organ systems to the postpubertal effects of testosterone secretion. The perinatal surge in testosterone also plays a role in inducing males to display male behavior postpubertally.

Measurement of testosterone in dogs and cats has not been systematically performed throughout the developmental period, but certain developmental periods have been sampled. Figure 23–2 shows a graph of the pattern of testosterone secretion and indicates that the perinatal surge in testosterone secretion is just before birth in dogs (Hart and Ladewig, 1979b).

During the later prenatal period, the gonads of males secrete testosterone, whereas the gonads of females are relatively quiescent in terms of hormone production. Testosterone in the developing male fetus has an apparent effect on its brain, and this makes it likely that the animal will display typical male behavior once it reaches puberty, and the brain is again acted upon by the pubertal secretion of testosterone. Lack of testosterone secretion perinatally tends to promote the development of female behavior, so that when the ovaries become active at the time of puberty, the secretion of estrogen activates the patterns of sexual receptivity.

Once the period of perinatal brain development has passed, the basic tendency toward masculinity or femininity seems to be permanently built into the brain. In fact, recent evidence in rodents illustrates that there are clear differences in the microscopic structure of some parts of the hypothalamus in males and females (Gorski, *et al.*, 1980).

Both males and females appear to have the basic neurologic elements for behavioral patterns typical of both sexes. The difference in brain development relates to the ease with which either the male or female system is activated. In males, normally, the secretion of testosterone at puberty tends to lead to the patterns of urine marking, to enhance the sexual interest of males to females, and to increase the likelihood of engagement in fights with other males.

Behavioral differences between males and females are evident in the juvenile stage of life even before puberty and before there is an appreciable difference in gonadal hormone secretion between the two sexes. For example, play behavior between male dogs includes much more mounting than in females. The urination posture differs with females using a squat posture and juvenile males using a stand-lean posture. These differences in urination posture occur even though the males may not be engaging in actual urine marking behavior prepubertally. Most differences between males and females, however, are accentuated at the time of puberty when there are pronounced differences in gonadal hormone secretion between the two sexes.

From the foregoing discussion, it is apparent that the neurological basis for male behavior exists from the time of birth. Pubertal androgen secretion normally activates these neurological systems, but it is clear that they may also be activated by environmental disturbances or social factors.

This can be seen in the occurrence of fighting in prepubertally castrated male dogs. The interaction with other animals, in combination with certain environmental factors, may activate a neural mechanism that is not otherwise active without testosterone stimulation.

In cats, the onset of the feline breeding season with all the attendant sociosexual furor, or the changing of homes, are examples of such environmental or social disturbances. Although most prepubertally castrated males never engage in spraying or fighting with other males, enough of them do so later in life to make this the most frequent behavioral problem in cats.

Normal females have a much lower tendency to show male behavior than castrated males. However, since they do have some capacity for male behavior, it is not surprising that female cats sometimes become problem sprayers, or female dogs start mounting people or getting into fights to an excessive degree. One explanation, of course, is that the same environmental and social factors that occasionally arouse male behavior in castrated males, may activate male behavior in females.

There is one other consideration with regard to masculine behavior in females. A type of masculinization can come about by the juxtaposition of a female fetus between two male fetuses in the uterus, so that there is diffusion of androgen from the male fetuses to the female fetus in sufficient quantity to cause a degree of behavioral masculinization. This effect has been well-documented in rats and mice (Clemens, 1974; vom Saal, 1981).

One clinical question that arises is whether the "womb-mate" effect would explain the occurrence of problem behavior in some females. It is known that female dogs can be masculinized by injecting pregnant females with testosterone. This notion has been pursued with spraying by female cats. A recent survey revealed that female cats that engage in occasional urine spraying have the same probability of coming from all female litters in which prenatal androgenization is unlikely, as from a litter with predominantly males in which androgenization is more likely (Hart and Cooper, 1984). Thus, the intrauterine masculinization effect appears to have no influence in cats, although it may play a role in dogs.

Finally, we should recognize that castrated males will have an increased tendency to display male behavior when they are given anabolic steroidal preparations for medical reasons. Androgens given to increase an aging male's general well-being and physical condition, or to ameliorate certain dermatological conditions, can activate the neurological systems of male behavior in the same way as postpubertal secretion of testosterone by the testes.

EFFECTS OF CASTRATION

Androgens that are produced in adulthood have a number of morphological, physiological, and behavioral influences. They maintain anabolic processes in some major muscle groups, and also effect metabolic rates and liver enzymes. Castration influences these processes.

Castration of Adult Males

In adult male dogs and cats the behavioral patterns that are altered by castration are those that are markedly sexually dimorphic. In dogs this includes the problem areas of aggression toward other males, urine marking in the house, mounting of other dogs or people, and roaming.

In cats the sexually dimorphic behavioral patterns of clinical interest are fighting (generally with other males), urine spraying in the house, and roaming.

Castration has an effect on the copulatory activity of dogs and cats. The studies on copulatory behavior have documented, under controlled laboratory conditions, one of the important principles in understanding the use of therapeutic castration for clinical problems. This principle is that a behavior is considered to respond to castration if castration reduces it in a significant proportion of the individuals of a species. Some individuals stop mating almost immediately after castration while others continue mating for years about as frequently as before castration. The differences cannot be related to age or sexual experience, but probably reflect differences in genetic makeup.

The same principles apply to clinical problem behaviors. Figure 23–3 illustrates the results of two clinical surveys on the effects of castration performed for problem behavior in male dogs (Hopkins, *et al.*, 1976) and male cats (Hart and Barrett, 1973). The graph illustrates one aspect that is also evident from the laboratory studies on mating behavior: castration is not as effective in altering behavior in dogs as it is in cats.

In cats one can expect intermale aggression, urine marking (spraying), and roaming to be greatly reduced or eliminated 80 to 90% of the time, with one half of the cases involving aggressive behavior and roaming, and almost all cases involving spraying, showing a rapid decline. In dogs, overall, intermale aggressive behavior, urine marking in the house, and mounting were greatly reduced or eliminated in only about 50 to 60% of the dogs studied. In about one half of those in which castration was effective, the decline in behavior was rapid; in the other one half, it was gradual. Some comments about the behaviors that are altered in dogs are in order.

Mounting in Dogs. One of the indications for castration is mounting of other male dogs or people. Some males persistently engage in mounting. Puppies often go through a phase of mounting of other dogs, people, or inanimate objects, and this mounting usually ceases as the puppies begin to mature. At other times mounting becomes quite pronounced at the time of puberty or sometime thereafter. The latter case would seem to be the most amenable to the effects of castration.

Urine Marking by Dogs. This is a problem when it occurs in the house. During or after puberty male dogs begin to assume the leg-lift posture and frequently mark vertical objects in their environment, apparently as an expression of territorial identification. On occasion, the stimulation from other dogs or some aspects of the environment may be such as to induce a male dog to mark in the house, even though he may be well-housebroken in general. This behavior may even be displayed in the presence of the owner and may not be eliminated even when severe punishment is administered. Castration often reduces urine marking in

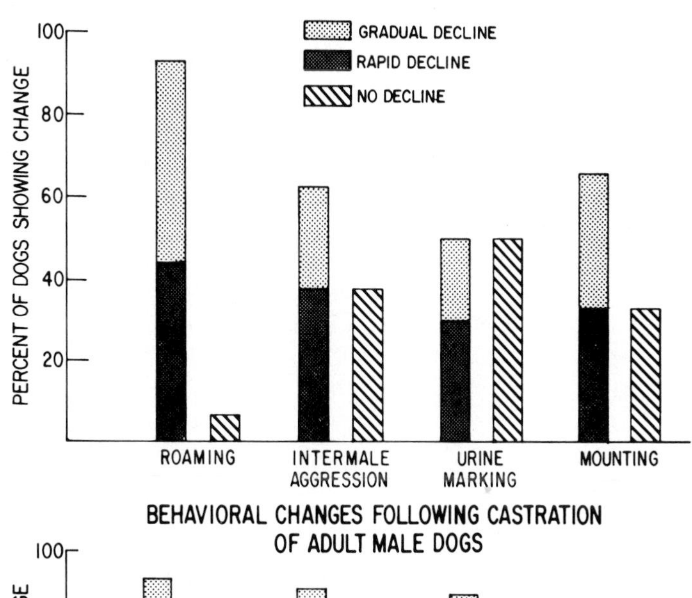

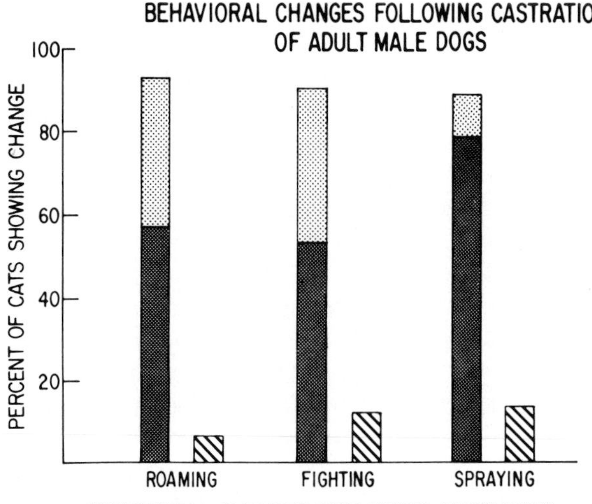

Fig. 23–3. This graph illustrates that castration more frequently alters problem behaviors in cats than in dogs. (From Hart and Barrett, 1973, and Hopkins, Schubert and Hart, 1976; reprinted with permission, American Veterinary Medical Association.)

the house but not outdoors. This difference is probably related to the fact that olfactory stimuli from the urine of other dogs, which are the most important stimuli in evoking urine marking, are strongest outside the home.

Aggressive Behavior in Dogs. There are several types of aggressive behavior. The type of aggression that appears to be the most amenable to alteration by castration is fighting between males. This can be a problem for people who have two male dogs and for individuals who take their male dogs into public places where other males are encountered. Because of their highly developed social communication, intermale aggessive behavior in dogs is not a serious problem as it is in cats. That is, dogs have a much greater tendency to settle agonistic encounters through threats, with one animal assuming a subordinate and the other a dominant role rather than by outright fighting. Before deciding on whether castration would be indicated for a type of aggressive behavior in dogs, it is important to determine if intermale fighting is the best diagnosis of the problem (see Chapter 3). Aggressive behavior between male dogs is altered in about 60% of the cases, with half the animals showing a rapid decline. One should use caution when suggesting a therapeutic approach to aggressive behavior because there are several types of aggressive responses as mentioned above. It was found that fear-induced aggressive behavior or fear biting were not reduced by castration.

Roaming in Dogs. If a male dog is not roaming for obvious nonsexual reasons, such as to visit a children's playground or to receive some food handouts from a distant neighbor, then castration is clearly indicated. The rate of effectiveness, about 90%, is higher than that reported for mounting, urine marking, and intermale fighting, which ranges between 50 and 60% in effectiveness.

The survey indicates that male dogs may be less active and more easily controlled after castration. A few owners in the survey noted that their dogs appeared calmer after the operation; others reported that their dogs became more affectionate and better with children. The changes in activity level and interaction with people are more difficult to document than changes in urine marking, mounting, and aggression toward dogs. Nonetheless, the changes were evident. In Chapter 21 it was noted that male dogs were ranked by experts as being more dominant over the owner, generally more active, and stronger on territorial aggression than female dogs. These traits would, by definition, be sexually dimorphic and one could expect a reduction in these behaviors in some dogs after castration. Such changes would be interpreted in different ways by the dog's owners. For some it might mean getting along better with children, for others it would be acting more mellow, or being more easily handled.

There is little evidence for alteration of behavioral patterns that are not sexually dimorphic. Therefore, generally desirable characteristics such as hunting ability and watchdog behavior are not altered. Similarly, problem behaviors for which there is little sexual dimorphism, such as bird-catching in cats, are not altered by castration.

There is no evidence that older animals, with presumably more experience in performing problem behaviors, are affected any less frequently than the younger, less experienced males. Therefore, the prognosis for the use of castration for indicated problem behaviors should probably not be affected by the animal's age or experience in performing the behavior.

At the present time it is difficult, if not impossible, to predict in which animals the operation is going to be most successful. There are undoubtedly genetic influences. Also, individual differences might be related to differences in the home environments of the animals.

The persistence of undesirable masculine traits following castration is apparently not due to any residual amounts of testosterone, since testosterone blood concentrations fall to low or practically nondetectable levels within 8 hours after castration.

Prepubertal Castration

It is common practice for male cats to be castrated before puberty. There is an assumption among veterinarians that prepubertal castration is more effective in preventing problem behaviors, such as urine spraying and fighting, than postpubertal castration in eliminating these behaviors once they have begun. However, it is not uncommon to find prepubertally castrated animals displaying objectionable male behavior. This is especially evident in cats where males, neutered at six months of age, begin spraying as late as three or four years of age. The onset of spraying is often related to the owners introducing new cats into a household with other cats, changing households, or altering a major aspect of the cat's lifestyle, such as making an outdoor cat an indoor cat.

The question arises as to whether prepubertal castration is actually more effective in preventing male problem behavior than postpubertal castration is in eliminating the behavior. Another concern is whether neutered males are more or less likely to engage in problem behavior than females, and also whether the age of the animal at the time of prepubertal gonadectomy is related to the likelihood of spraying.

The most important clinical areas in which this issue arises are urine spraying and fighting in cats. This issue was addressed in a recent survey of 134 male and 152 female cats gonadectomized between 6 to 10 months of age (Hart and Cooper, 1984). The study found no relationship between the age of male cats at the time of prepubertal castration and the likelihood of spraying or fighting. Also, there was no relationship between the age of ovariohysterectomy in female cats and the occurrence of these behaviors.

Frequent urine spraying by prepubertally castrated male cats was 10%, compared with 5% in prepubertally spayed females. Inasmuch as only 10% of male cats castrated for spraying in adulthood continue to spray, this survey revealed that prepubertal castration is not likely to be more effective in preventing objectionable spraying than postpubertal castration is in eliminating the behavior once it has started.

Although such an extensive study has not been conducted on dogs, the information available is along the same lines. Male dogs castrated at

40 days of age and tested as adults displayed aggressive behavior and mounted females as often as normal control dogs (LeBoeuf, 1970). These observations on cats and dogs, pointing out that prepubertal castration does not impair masculine behavior any more than castration in adulthood, are paralleled by similar findings on sexual behavior in laboratory rodents.

It is probably true that prepubertal castration is not qualitatively different from postpubertal castration, and the main influence, if there is any, is the time gained by castrating the animal early. That is, since castration often results in a gradual decline in various aspects of male behavior, castrating animals prepubertally simply extends the time for the effect of castration to occur. Perhaps our concept of the profound effects of prepubertal castration stems from the obvious effect it has on the body size and shape of males, and we erroneously assume that it will have an obvious effect on their behavior.

The data revealing a lack of correlation in the age of castration and the occurrence of spraying and fighting in cats may be of value in counseling cat owners who may be concerned about the possible effects of early castration in prediposing male cats to urinary blockage, or who wish to allow their males to grow the larger head, heavier jowls, and general morphology characteristic of tomcats.

BEHAVIORAL EFFECTS OF OVARIAN HORMONES

The ovarian hormones, estrogen and progesterone, are of primary concern in the behavior of the females. Estrogen is produced by the developing Graafian follicle and reaches its peak in secretion just before the onset of estrus. As far as is known, in all mammalian species (except the human female), an increase in estrogen secretion is needed to bring on a full display of sexual behaviors. In some species the secretion of progesterone seems to facilitate the display of behavioral estrus. The secretion of estrogen is controlled principally by the gonadotropins from the anterior pituitary.

Changes in sexual behavior brought about by the secretion of estrogen are increased attractiveness to males, initiation of solicitous responses such as seeking out males (if allowed the freedom), and the display of receptive behavior in response to a male's sexual advances. Other changes not directly related to copulation are an increase in general activity and an increase in urination frequency in dogs.

The concept of sexual dimorphism and the role of androgen during the development of the brain have been dealt with in the discussion of male hormones and behavior. As far as is known, there is no behaviorally significant amount of ovarian hormone secretion during early life. This lack of secretion of any sex hormone leads to the development of a nervous system along feminine lines. In fact, the administration of fairly large doses of estrogen to neonatal female rats actually suppresses the full development of a feminine system.

Effects of Estrogen in Adult Females

In studies of the hormones necessary for the production of estrous behavior in spayed females, it has been found in all species that estrogen alone will induce receptivity if enough of the hormone is given.

Female dogs ordinarily go through one estrous period twice a year, and there is not a pronounced seasonal effect on the occurrence of estrus. In female dogs and cats, as in most other mammals, estrogen increases general activity. Females during estrus usually move about more, vocalize more frequently, and may act somewhat nervous. The urine and vaginal secretions of females in proestrus are attractive to males. The attractants are undoubtedly noticed by males living in the vicinity of females or by males on their daily treks through the females' vicinity. It is not known whether the attractants are metabolites of estrogen, or a secretion added to the urine and vaginal secretions when estrogen secretion is high.

There are stories about male dogs being attracted from miles away to the vicinity of an estrous female, presumably by one or more of the sex attractants in the urine. These attractants are probably not so potent as to be detected by males from miles away. Some female dogs display the masculine leg-lift urination posture when they are in estrus. The function of this behavior is probably to assure that the urine is deposited on prominent vertical objects in the environment.

Estrous behavior in cats can be easily evoked without progesterone, and there are no reports of progesterone facilitation. In dogs, estrogen alone will evoke sexual receptivity, but an injection of progestone the day before, or the day of, estrus seems to have a facilitatory effect with respect to the female's attractiveness to the male (Beach and Merari, 1970).

Just as estrogen can evoke sexual responsiveness in castrated males, testosterone tends to mimic estrogen in evoking receptive behavior in the female. This has been documented in laboratory rodents.

Effects of Progesterone

With the exception of the special facilitatory effects on induction of estrus in some species, progesterone has an inhibitory effect on female sexual responsiveness and it has some suppressive effects on male behavior as well. In laboratory rodents progesterone has been shown to inhibit male sexual, aggressive, and scent-marking behaviors. These effects occur apart from the progesterone suppression of gonadotropin secretion. It is noteworthy then that progesterone has effects on behavior opposite to those of both testosterone and estrogen. In large doses, progesterone produces general anesthesia in various animals, and there are indications that the hormone has calming or tranquilizing effects in somewhat lower doses.

Some of the observations regarding the behavioral effects of progesterone in women are interesting. It is estimated that near the end of pregnancy in humans, progesterone secretion is ten times that obtained during the peak of progesterone secretion during the menstrual cycle (Hamburg, Moos, and Yalom, 1968). This is believed to be one of the reasons women experience a general feeling of well-being and calmness during pregnancy. The dramatic fall in progesterone at parturition is believed to precipitate depression and aggressive tendencies in a few women. Since progesterone falls just before menstruation there has been some interest in determining if the fluctuations in progesterone levels are related to behavioral and mood changes in women. Scientific opinion has gone back and forth over this issue.

Effects of Ovariectomy

At the time of puberty, the ovaries of normal females start secreting estrogen and progesterone in a pattern characteristic of the particular species. However, if the female is spayed during or after an estrous period, sexual behavior will not be displayed again. Females are thus considerably different from males which continue to show sexual behavior for varying periods of time after gonadectomy. There are no reports of any enduring behavioral differences resulting from gonadectomy performed before or after the first estrus.

Experiments on rats have documented that ovariectomy leads to increased appetite, increased food intake, and decreased general activity. In female dogs the gain in weight from ovariohysterectomy was found to be 5 to 10% in one study in which controlled observations were made on gonadally intact females (Houpt, *et al.,* 1979). This is much less than that reported for rats, but following ovariectomy, the gain in body weight from enhanced food intake and reduction of general activity would logically be greatest in females that cycle throughout the year, such as rats.

Ovariohysterectomy is, of course, the most common way of removing the ovaries and the effects of estrogen secretion. Essentially this produces a constant state of diestrous. From this standpoint a gonadectomy does not make a female dog less feminine in the way that it might be thought to reduce or eliminate the masculine behavioral characteristics of males. There seems to be no justification for not removing ovaries in a female on the basis that she might be more feminine if allowed to cycle occasionally, since we generally consider female cats and dogs just as feminine when they are in diestrus as when they are in estrus.

What about the timing of an ovariectomy with regard to behavioral consequences? There is no evidence to show that allowing a female to go through one or more estrus cycles has any enduring effect on behavior. The same line of thinking applies to any expected behavioral changes developing from a bitch having experienced pregnancy and the nursing of puppies. Obviously, the behavior of the animal is different while under the hormonal influences of ovarian and pituitary hormones, but there is no clinical or experimental evidence that the changes are of an enduring nature.

BEHAVIORAL EFFECTS OF EXOGENOUS PROGESTINS

The single most useful type of drug for behavioral therapy in small animal practice is the synthetic progestin. At the present time, none of the progestins are specifically marketed for use in the behavioral problems of animals. Through the efforts of practitioners to find a treatment for objectionable urine spraying in cats, synthetic progestins have gained prominence as behaviorally effective drugs. Gerber and Sulman (1964) were the first to report that a synthetic progestin would suppress malelike behavior. At the present time both the injectable medroxyprogesterone acetate Depo-Provera [Upjohn] and megestrol acetate Ovaban [Schering] have been found to have suppressive effects on objectionable masculine behavior. The reason the long-acting progestins are used rather than progesterone in oil is that progesterone is metabolized within a few hours and one would have to administer it every few hours for it to have any therapeutic effect.

The mechanism by which the above behavioral patterns are altered is not understood. Theoretically, progestins mimic the action of the hormone progesterone, except that they sustain effective systemic blood levels of the active agent for weeks or months rather than a few hours. The best current explanation is that the progestins seem to have an inhibitory effect on brain mechanisms mediating male behavior as well as a calming and tranquilizing effect. We should not expect behavioral problems that fall outside the categories discussed above, such as fear-related aggression or bird catching, to be altered by progestins.

Medroxyprogesterone acetate and megestrol acetate are derivatives of 17-α-acetoxyprogesterone. Practitioners are familiar with medroxyprogesterone acetate as an injectable drug and megestrol acetate in tablet form and tend to use one or the other according to past experience or the wishes of the client. It has been found that megestrol acetate is sometimes effective when medroxyprogesterone acetate was not and vice versa. In particularly severe problems, practitioners may inject medroxyprogesterone acetate and dispense megestrol acetate tablets as a supplement.

The dosage of either drug has never been determined by customary pharmacological experimentation, undoubtedly because it is practically impossible to duplicate the behavioral problems of cats in the laboratory. Hence, the dosage reflects a clinical "trial-and-error" approach. Since the behavioral problems we are concerned with are quite severe and disruptive of the pet-owner relationship, dosages have tended to be on the high side. Medroxyprogesterone acetate is an aqueous suspension and is usually injected subcutaneously or intramuscularly at a rate ranging from 10 to 20 mg/kg for cats and 5 to 10 mg/kg for dogs. Megestrol acetate is given orally in tablet form at the rate of 5 mg per day to cats. If medroxyprogesterone acetate is going to be effective, it will be clear with the first injection. An additional injection is given if needed, no sooner than one month.

Blood levels of medroxyprogesterone acetate reach high levels for the first 20 days, then plateau before beginning a gentle downslope. Seventy-five days after an injection, levels of medroxyprogesterone acetate are one third of those found after 7 days.

For megestrol acetate treatment, it is usually recommended that the owner administer the tablets once a day for about one week to determine if the behavior is altered. If it is, then the megestrol acetate tablets should be given once every two days for the next two weeks, and if the behavior is still controlled, the dosage should be phased down to about once per week. Megestrol should be given for probably no longer than 4 to 6 months.

For both progestins, recent analyses indicate that one can expect a significant improvement in problem behavior for which the drugs are indicated in no more than 50% of the animals treated (see Section II).

The specific effectiveness of progestin therapy has been extensively evaluated in cats and to a more limited degree in dogs. In cats the progestins are used for urine spraying (see Chapter 14) or marking and aggressive behavior. Both medroxyprogesterone acetate and megestrol acetate were almost equally effective in the initial treatment of spraying. However, as many as one half of the cats not responding to medroxyprogesterone acetate treatment were found to respond favorably to subsequent megestrol acetate treatment. The effectiveness of progestins for aggressive behavior in cats (about 75% of cats treated) is higher than for treating spraying (Hart, 1980; unpublished). In dogs the progestins have been successfully used to treat aggressive behavior, urine marking in the house, and mounting of people.

Data are available for aggressive behavior in which it was found that treatment with medroxyprogesterone acetate was effective in markedly reducing or eliminating fighting with other males in 75% of dogs. Only 1 of the 7 dogs (a castrated dog) treated for aggression toward the owner responded favorably. The only side effects observed were an increase in appetite and weight gain. These occurred in about one third of the dogs (Hart, 1981a).

At one time, medroxyprogesterone was used to prevent estrus in bitches. However, because of its tendency to cause endometritis and pyometra, presumably due to overdosage or use at the wrong time in the estrus cycle (Stabenfeldt, 1974), it was withdrawn from the market as a canine contraceptive. In the doses used for behavioral therapy, medroxyprogesterone acetate can be expected to disrupt normal testicular function and to cause sterility in males. Therefore, the drug would be contraindicated in breeding males.

The individual variability in the effectiveness of these drugs is probably a reflection of genetic differences among animals, just as the effects of castration seem to be a reflection of genetic differences. Some individual differences are also undoubtedly related to the type of predisposing factors that may bring on a problem, and these vary greatly from household to household. For example, the presence of female cats seems to predispose male cats to spraying. It is usually easier, therefore, to control spraying in a cat that is the only feline in the household than one that shares its house with a female. One should keep in mind that, at least for male behavioral patterns, neutered males and females have a relatively low tendency to show the behavior; when the behavior does occur, it is probably the reflection of disturbing environmental factors which are severe enough from the animal's standpoint to cause the behavior.

Undesirable Side Effects

The most frequently reported side effect is an increase in appetite. This occurs in approximately one third of the cases. Often the animals gain considerable weight. In about one fifth of the cases, the owners report their animals seemed depressed, lethargic or inactive, at least for a short time after drug administration. Enlargement of the mammary gland has been noted in a few (6%) of the cats treated. In no case did tumors develop (Hart, 1980c). In Beagle dogs medroxyprogesterone acetate treatment has resulted in an increase in breast nodules with some nodules metastasizing after approximately 3 years of treatment (Frank, *et al.*, 1979).

It is somewhat instructive to examine the occurrence of these side effects against the background of clinical reports from the use of medroxyprogesterone acetate in human medicine. Medroxyprogesterone acetate has been used for a variety of conditions, including treatment of endometriosis, threatened and habitual abortion, and uterine cancer. The drug has been explored for use as an injectable contraceptive for both women and men. It has also been used to slow precocious puberty in children. The only reported behavioral use in humans for which the drug has been used is for the treatment of male sex offenders.

The doses for these varieties of human uses have varied considerably. A dose as low as 150 mg will prevent ovulation if injected every 3 months intramuscularly. Doses have ranged upward to 1200 mg monthly for treatment of cancer or with trials for use of the drug as a male contraceptive.

Judging by the dosages that have been employed in the human field, the amounts of progestins customarily given to dogs and cats must be considered very high. These doses have been found to have a suppressive effect on the adrenal cortex, reducing plasma corticosteroid levels by 75% and lowering ACTH levels as well (Chastin, Graham, and Nichols, 1981).

The progestin medroxyprogesterone acetate has also been found to result in diabetogenic stress of moderate magnitude as indicated by high blood glucose. The high doses in people have resulted in a worsening of the diabetic condition and may precipitate a diabetes syndrome in prediabetic patients. Another common side effect in human patients administered medroxyprogesterone acetate is an increase in appetite and subsequent weight gain of 2 to 10 pounds.

Obviously, the progestins must be administered with the potential side effects kept in mind. Those that are most serious, including the possible precipitation of diabetes mellitus, depression of corticosteroid output, corticosteroidlike effects, and production of mammary gland abnormalities, appear to result from long-term administration at a fairly high dosage. The less serious side effects, namely appetite stimulation or suppression and occasional depression and inactivity, are often observed shortly after initial administration of either drug.

It would appear that medroxyprogesterone acetate and megestrol acetate are most justifiably used in situations in which one or two treatments produce a satisfactory resolution of the problem. However, the behavioral problems for which the progestins are indicated are often so serious that the owners are considering euthanasia as an alternative. Thus, even the risks of adverse side effects from very prolonged treatment may be entirely justified.

Certainly the periodic evaluation of the health of a dog or cat under long-term progestin therapy is advisable. This should include examination of glucose levels in the urine and frequent palpation of the mammary gland region.

Chapter 24

Psychoactive Drugs and Behavioral Therapy

Drugs have revolutionized the treatment of human psychiatric patients in mental hospitals and have made possible the resolution of some major mental disorders without hospitalization. Just as important, psychoactive drugs have allowed normal people to deal with severe psychological crises that might otherwise have proven socially or culturally disabling. Why has not the same progress been made in the area of clinical animal behavior? The use of drugs for behavioral disorders or problems in animals is conceptually quite different from their use in people. In people, much of the emphasis on the success or failure of drug treatment is based on verbal reports. In animals, one can rely only on behavioral signs for indications of drug effectiveness. Animal owners are also more tolerant of moderate behavioral disturbances in animals than in people. In this chapter we will examine the basic principles and information relevant to the use of drugs in behavioral therapy. It should be emphasized from the outset that it is important to pursue some form of behavioral therapy while treating the animal with a drug.

Aside from differences in clinical approaches to problem behavior, there appear to be basic differences between animals and people in their behavioral reactions to drugs. Some differences may simply reflect the fact that animal behavior includes a greater predominance of genetically determined responses than human behavior, and some drugs may influence innate behavioral patterns in different ways than acquired behavior. A case has been built for ascribing some differences in behavioral responses in man and animals to species differences in the biotransformation of drugs in the central nervous systems (Brodie, 1962). By the same token there may be important differences in drug action between different animal species.

The sedating action of tranquilizers has led to the rather widespread use of these drugs for the purpose of restraining animals. A tranquilized animal can be more easily handled during a physical examination, held for X-ray, given medication, or transported in automobiles or planes.

Some tranquilizers have also proven quite valuable as preanesthetic agents. The correction or treatment of behavioral problems in animals implies a different approach than restraint or preanesthesia because specific behavioral problems can require different drugs.

Although the available information to guide drug use in animals is very limited, animals have been used in experimental studies of psychoactive drugs. Some of these experimental reports are of value in predicting the usefulness of a drug in treating a particular behavioral condition. The trouble with such information is that the behavior is evoked by unnatural behavioral conditioning procedures or brain lesions, and these conditions may bear little resemblance to behavioral problems that occur spontaneously in animals.

OVERVIEW OF DRUGS USED IN PSYCHOPHARMACOLOGY

Psychopharmacological agents including tranquilizers, stimulants, and antidepressants are often referred to as psychoactive or psychotropic drugs. Substances that cause alteration of normal psychological states, such as lysergic acid diethylamide (LSD) or mescaline, are referred to as psychogenic or psychedelic drugs. Before the advent of psychoactive drugs for the treatment of severe mental disorders in man, psychotherapeutic approaches left much to be desired (Himwich, 1960). Sometimes patients exhibiting extreme hyperactivity could be controlled only by stupefying doses of barbiturates. Improvement in patients with severe depression or extreme hyperactivity was sometimes obtained with the use of electroshock or electroconvulsive therapy. This procedure is still in use today, but to a more limited extent. It involves passing a brief but intense electric current through the brain with electrodes placed on the temples. The physiological process by which the improvement is obtained is not understood, but there is probably no similarity between the mechanisms by which electroconvulsive therapy and drugs alter behavior. Electroconvulsive therapy has been used with some reported success with aggressive behavior in dogs (Reddy, 1978). Another type of shock treatment involves the administration of a dose of insulin sufficient to reduce the glucose levels in the blood to the point at which the brain goes into a coma. In some instances in which electroshock therapy is not effective, insulin shock may prove to be of value.

Tranquilizers

Tranquilizers, or ataratics, comprise a large array of psychoactive substances. Reserpine was the first potent tranquilizer developed. Although this drug is still used in research, it is no longer used for clinical purposes. Chlorpromazine was produced soon after reserpine, and it is still widely used clinically. Chlorpromazine is related in chemical structure to the largest group of phenothiazine derivatives, which is the most widely used group of tranquilizers in canine and feline medicine.

Some tranquilizers are effective in treating human psychoses such as schizophrenia or paranoia. Psychoses are behavioral disorders so severe as to incapacitate an individual's functioning in society. A major characteristic of the psychotic condition is an impairment of thought processes. Tranquilizers that have specific antipsychotic powers in people are sometimes classified as neuroleptics or major tranquilizers. These include most of the phenothiazines. The so-called minor tranquilizers, which include the benzodiazepine derivatives, are ineffective in treating the cognitive disturbances of psychotic patients, but are effective in treating certain neurotic conditions such as excessive anxiety or nervousness in people with essentially normal thought processes.

There is some disagreement as to whether the neuroses and psychoses in people are fundamentally similar forms of the same type of behavioral disorder and differ only in respect to severity, or whether they are qualitatively different types of behavioral disorders (Vandenberg, 1960). The observation that only some drugs are effective in treating psychotic patients argues for the latter hypothesis. The neurosis-psychosis dichotomy applies mostly to human psychopharmacology, since there has been no documentation in animal analogs of the two general conditions. There are some apparently inherited conditions, such as idiopathic aggressive behavior and flank-sucking in dogs, and wool-chewing in cats, which one could classify as psychotic behavior. However, these behavioral disturbances do not generally respond to phenothiazine or other tranquilizer treatments.

Tranquilizers have calming properties, and usually result in a reduction of spontaneous activity and a decrease in response to external or social stimuli. Although the major tranquilizers, such as chlorpromazine, have been successfully used to alleviate anxiety in neurotic and normal human patients, the minor tranquilizers seem to be favored because the sedative, or other similar side effects of these drugs, are minimized, thus allowing the individual to work as usual.

The following is a discussion of the different types of tranquilizers available for human and animal use.

Phenothiazines. This is the major group of tranquilizing drugs that has been used in domestic animals. The group includes chlorpromazine (Thorazine), promazine (Sparine), acetylpromazine (Acepromazine), perphenazine (Trilafon), trimeprazine tartrate (Temaril-P), propiopromazine hydrochloride (Tranvet), triflupromazine (Vetame) and piperacetazine (Psymod). For purposes of pharmacological restraint, or as a preanesthetic, these drugs are usually given intravenously or intramuscularly. There are some species differences in reaction to these drugs which are important to consider especially when they are administered intravenously.

Chlorpromazine, acepromazine, and piperacetazine, the most commonly used drugs for pharmacological restraint, are used to treat some behavioral problems in dogs and cats. The major side effects of long-term use are cardiovascular disturbances, especially hypertension and extrapyramidal signs such as ataxia, muscle tremors, and uncoordination. Promazine causes a greater degree of hypertension than any of the others (Jarvik, 1970) and is probably avoided for this reason.

Benzodiazepines. Diazepam (Valium) and chlordiazepoxide (Librium) are not officially approved for use in dogs and cats but the drugs, especially diazepam, are widely used. They have less cardiovascular and sedative side effects than the phenothiazines. In animal use, it is not uncommon for diazepam to produce some degree of ataxia at high dosages. Diazepam is considered to be the more potent of the two drugs.

Butyrophenones. Haloperidol (Haldol), droperidol (Inapsine), and azaperone (Stresnil), have been used in human patients. The use of these drugs in animals has been explored to only a limited degree. Azaperone has been found to reduce aggressive tendencies in pigs (Symoens, 1969). These drugs could be used in some animals as an alternative when the other groups appear to be ineffective.

Meprobamate. This drug, basically one of a kind, has been used in the past to treat neurotic conditions, excessive anxiety, and tension in people. Meprobamate has been replaced almost completely by diazepam in human medicine. In animal studies meprobamate has usually proven less effective than other drugs for a variety of behavioral conditions.

Lithium Carbonate

In human patients lithium has been found to be quite effective in suppressing the manic phase of manic-depressive psychosis. For this reason it is referred to as a mood stabilizer. The main advantage of lithium over chlorpromazine, which has also been used to treat manic episodes, is that lithium does not have sedative characteristics. Extensive use of lithium in animals is probably unrealistic because it is toxic in excessive amounts, and it is necessary to regularly monitor blood concentrations of individuals taking the drug.

Anticonvulsant Drugs

These drugs including phenytoin (diphenylhydantoin) (Dilantin) and primidone (Mylepsin) are routinely used to control animal and human seizures associated with epilepsy. The drugs, especially phenytoin, have been used to control hyperactivity, explosive aggression, and irritability in a variety of human patients (Resnick, 1967). Some small animal practitioners have found that this drug is useful in suppressing the idiopathic aggressive behavior of dogs. Phenytoin and primidone are somewhat toxic for cats and should not be used in that species.

Progestins

This is the one group of hormonelike drugs that has general value for behavioral therapy in small animal practice. Frequently used drugs are the injectable medroxyprogesterone acetate (Depo-Provera) and the oral form of megestrol acetate (Ovaban). In addition to suppressing undesirable male behavior, the synthetic progestins have calming effects, not unlike that of tranquilizing agents. The effects and indications of progestins are discussed in detail in Chapter 23. Interestingly, this is the one group of drugs that has wider application for problem behavior in animals than people.

Stimulants

Central nervous system stimulants such as amphetamine and methylphenidate (Ritalin) have been used with some success in treating depression in human patients. An interesting use of the drug methylphenidate is in treating children with a hyperkinetic or hyperactive condition. The effect is considered somewhat paradoxical because such children are actually calmed by this drug whereas normal children are activated. Stimulants have been reported to reduce abnormal hyperactivity in dogs that cannot be controlled by behavioral techniques (Corson, *et al.*, 1977). The use of these stimulants for hyperactivity is indicated only when the physiological signs of excitement (heart rate and respiration rate) are slowed as well as the behavioral hyperactivity.

Antidepressants

These drugs are often more effective than tranquilizers in treating mood disorders in people with severe depression. There are currently two groups that have been found useful. The monoamine oxidase inhibitors have the ability to block oxidative deamination of brain amines such as dopamine, norepinephrine, epinephrine, and 5-hydroxytryptamine, and thus they raise the concentration of these substances in the brain. The monoamine oxidase inhibitors seem to stimulate a mood elevation, sometimes apparently to the point of euphoria (Jarvik, 1970). These drugs are used infrequently because they require dietary restrictions and they interact with other drugs.

The tricyclic antidepressants include imipramine (Tofranil) and amitryptiline (Elavil). This group is thought to relieve depression by the dulling of depressive ideation rather than by mood stimulation. Some drugs in this group have been useful in combination with tranquilizers.

In normal animals these drugs do not have marked influences on behavior (Jarvik, 1970) and there are no specific reports of the value of these drugs in treating behavioral disorders in animals. If one could identify behavioral depression in a dog or cat, it would be interesting to note the effectiveness of these drugs. There are no systematic surveys of such applications.

MECHANISMS OF DRUG ACTION

Psychoactive drugs influence behavior most directly through altering the activity of neurons in the brain. This apparently comes about through the action of drugs at synaptic membranes. Some drugs may influence the synapse because they are identical or very similar, structurally, to naturally occurring inhibitory and excitatory synaptic mediators. Drugs may also influence synaptic activities by attaching to the receptor sites and blocking normal activation of the receptor sites by naturally occurring synaptic mediators. Other drugs may influence the metabolic pathways involved in the synthesis of synaptic agents.

Drugs may influence the activity of neurons by the above mechanisms, but there would be limited behavioral effects if the influences were uniform throughout the entire central nervous system. It is probably because neural structures are differentially sensitive to the biochemical action of various drugs that most behavioral effects are produced. That is, under the influence of a given drug the relative activities of different parts of the brain are altered, and hence a particular behavioral state is induced by altering the balance of neural activity in different structures.

Some anesthetics have sedative or hypnotic effects because they depress the reticular activating system more than other parts of the brain. Stimulants, such as caffeine or amphetamine, excite the reticular activating system more than other parts of the brain. Some tranquilizers appear to affect the hypothalamus and parts of the limbic system more than parts of the brain such as the cerebral cortex or cerebellum.

Effects on Neurotransmitters

Different parts of the brain appear to have different thresholds to drugs because of differences in the concentrations of neurotransmitters in these structures. The monoamines, serotonin and norepinephrine, are most highly concentrated in the limbic system, hypothalamus, and midbrain. Thus, drugs altering the concentration of these substances will have the greatest effect on these brain areas. Chlorpromazine blocks receptor sites and thus prevents monoamines from stimulating other neurons. The monoamine oxidase inhibitors prevent the enzymatic breakdown of the monoamine transmitters. Imipramine prevents the return of released monoamine transmitters back into the mother cells. Both types of antidepressants thus prolong the action of monoamines, whereas chlorpromazine blocks the action of monoamines.

The major differences between the biochemical action of the so-called minor tranquilizers and the phenothiazines is that the former are found to have a narrower or more limited action in the structures associated with the limbic system (Himwich, 1965). By the same token, undesirable neurological side effects are fewer and milder with the minor tranquilizers.

The Concept of Neurochemical Lesions

Although it is an oversimplification, it is conventional to think of some behavioral disorders as reflecting the existence of a neurochemical lesion, and that a psychoactive drug may reverse the effects of the lesion by stimulating a brain area that will overcome the biochemical abnormality. The dramatic improvement of mental patients suffering from severe depression following administration of an antidepressant may be an example. In such patients the drug produces a virtual disappearance of symptoms after a two- or three-week period. Tranquilizers such as chlorpromazine may mask a behavioral disorder by producing a compensating lesion rather than normalizing the biochemical lesion as in the example involving the antidepressant. Thus, schizophrenic patients treated with chlorpromazine may show toxic effects that mask their behavioral symptoms (Hamilton, 1964). The analogy is drawn to the neurosurgical treatment of Parkinsonism. The disease is due to a naturally occurring lesion of the extrapyramidal system, and the neurosurgeon produces an artificial (presumably compensating) lesion in the extrapyramidal system to relieve the abnormal tremors of Parkinsonism. Although the tremors are gone, no one would suggest that the brain was now functioning normally. The administration of L-dopa presumably produces improvement in Parkinson patients by replenishing the deficiency of L-dopa, which initially produced the disease.

Alteration of the Internal Environment

Another general way in which drugs may influence behavior is by acting on peripheral organ systems, causing an alteration of the organism's internal environment such that sensory input to the brain from visceral organs is markedly changed from that which existed prior to administration of the drug. Some drugs obviously affect both peripheral structures and the brain. Epinephrine has stimulatory effects on some parts of the brain, and it has marked peripheral effects, such as increasing blood pressure, heart rate, and slowing intestinal peristaltic activity.

Consider an animal that is exhibiting an undesirable response to a certain stimulus. It is natural to concentrate on the obvious physical stimulus, forgetting that stimuli also involve the internal environmental fields such as temperature, movement of viscera, osmotic pressure, and blood chemical levels. If a drug, by acting on the autonomic nervous system, alters the internal environment markedly, then the total stimulus field may be sufficiently changed, so that a response previously evoked by the total stimulus field might be altered.

Individual Differences in Drug Effects

A number of experiments using precision control groups and exacting experimental procedures have revealed that within a species, factors such as hereditary differences, group composition, prior experience, diet, and even the time of day of drug administration may markedly influence an animal's response to a drug (Broadhurst, 1964; Rushton and Steinberg, 1964; Stroebel, 1968; Watson, 1964; Wilson and Mapes, 1964). It is clear that every clinical case represents a new experiment in determining the appropriate drug form and dosage.

PRINCIPLES OF DRUG USE

In contrast to the use of drugs for restraint and/or to reduce anxiety for travel purposes, in which only a temporary change is required, the goal for drug use in treating behavioral problems is a permanent change in behavior. The approach is to use drugs as a tool to alter behavior, such as in desensitization, so that the undesirable behavior remains altered when the drug is no longer given.

Whether a drug is used on a permanent or temporary basis, it is necessary to find an agent that most effectively alters the behavior. This will be dealt with in the final section of this chapter. In this section we will assume a drug is being employed which is effective in altering the undesirable behavior.

A calming effect is generally associated with tranquilizers. In fact, the term "tranquil" refers to a state of calmness. This concept is misleading since some drugs referred to as tranquilizers do not always produce a calming or tranquilizing effect. One of these drugs, diazepam, has a tendency to sometimes increase excitability in dogs, which is hardly a calming effect. On the other hand, diazepam is the drug of choice for reducing or "calming" fear reactions. Tranquilizers, if given in high doses, have sedative effects and obviously can have calming effects in this manner. Ideally, we should choose a psychoactive drug for specific behavioral reactions such as reducing aggression toward other dogs, eliminating fear of strange people, or suppressing excessive excitability.

The typical classification of drugs for human use is for antianxiety (benzodiazepines), antipsychosis (phenothiazines, butyrophenones), antidepression (tricyclic antidepressants, monoamine oxidase inhibitors), and mood stabilization (lithium carbonate), but these categories are not relevant to animal use. Antidepressants and lithium have not been reported to have any effect on the normal or problem behavior of dogs and cats. The effectiveness of so-called antianxiety versus antipsychotic drugs on dogs and cats does not relate to any differences in the problem behavior profiles.

Stimulus Desensitization

The concept of stimulus desensitization, and the importance of dealing with the stimulus on a gradual basis, are covered in Chapter 22. Desensitization of an emotional reaction may require the use of a drug because the emotional response itself may interfere with adaptation to the stimulus. It is almost impossible, for example, to approach a pet that is aggressive toward people if the aggression is fear-induced. Also, the arousal of emotional reactions with accompanying sympathetic activation may physiologically prevent desensitization. On the other hand, a drugged animal that is overly sedated may not be responsive enough to allow for desensitization. This is why tranquilizers rather than barbiturates are used.

The behavioral approach to desensitization is rather straightforward. An animal that is fearful of a particular stimulus is administered a drug, usually a tranquilizer, to the point at which it does not show the undesirable behavior. If the behavior is not altered, or pronounced side effects occur, a different drug is usually tried. Once the behavior is satisfactorily altered and the animal is on a continuous administration schedule, it is then repeatedly presented with the stimulus that previously produced the emotional reaction (Fig. 24–1). This may be an active process, such as presenting a sound or visual stimulus during daily training sessions, or a passive process, such as allowing the animal to live in an environment that previously aroused fear responses. This desensitizing process is continued while the animal is maintained on the drug for two to eight weeks.

When it is felt that substantial progress may have been made, the drug dose is reduced by one-fourth to one-third. If the animal still behaves satisfactorily at this reduced level, it is maintained at this point for another week or two before the dose is subsequently reduced again. If the undesirable behavior occurs, the drug is increased and lowered again only after further desensitization. If adaptation or desensitization occurs satisfactorily, then successive reductions of the drug at one- or two-week intervals allow the animal to adjust to the situation gradually until the drug is no longer administered.

Recall the concept that when a drug is administered to an animal, the behavior is altered because the internal environment is altered. It is imperative that when treatment begins, the stimulus change be maximal. Changing the internal environment gradually may allow the animal to display the undesirable response to a broader range of stimulus conditions. The behavior could then be even more difficult or impossible to alter later, even with marked alteration of the internal environment. Gradually raising the tranquilizer dose may simply allow the animal to slowly adapt to moderate changes in internal environmental stimuli.

Aggressive Behavior

One of the areas in which tranquilizers are indicated is to suppress aggressive behavior when it is necessary to introduce new animals to an

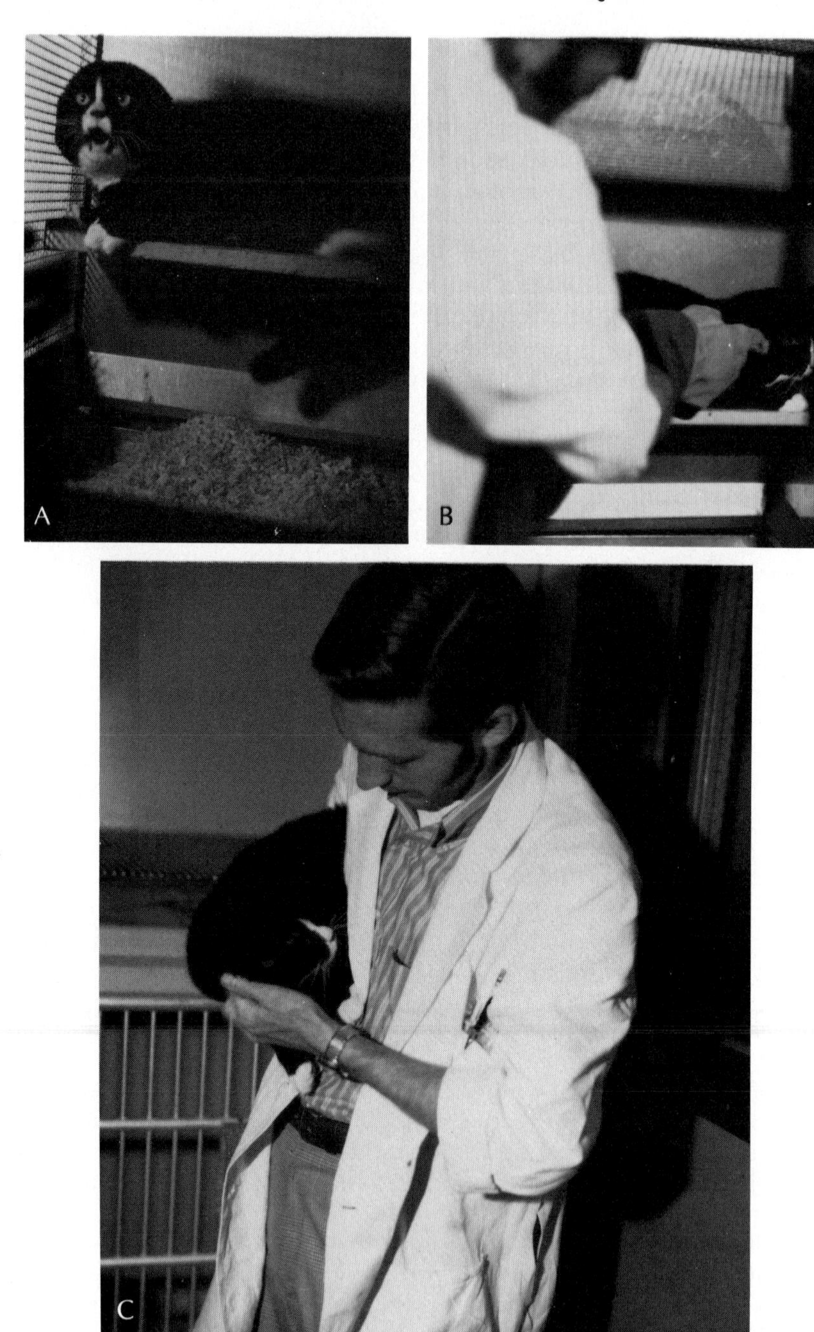

Fig. 24–1. Effects of a benzodiazepine drug (chlordiazepoxide) in reducing fear-related aggression in a cat: *(A)* patient in a nontranquilized state; *(B)* one day after tranquilizer treatment; *(C)* two days after treatment.

established group, or to place animals together to form a new group. This procedure has been documented quite well with pigs. Azaperone was given to pigs at the rate of 2 mg/kg prior to mixing strange animals together. The drug markedly reduced initial fighting, and the fights that did occur were less prolonged. In subsequent trials it was learned that with light doses (less than 1.5 mg/kg) the pigs would lie down for one-half to one hour but still fight violently when disturbed or startled. Moderate doses (1.5 to 3.0 mg/kg) resulted in sedation and even ataxia for approximately two hours, but after this, the animals walked around and seemed to pay little attention to each other. The occasional fights that did occur were not serious. At high doses (over 3 mg/kg), deep sedation occurred and fighting later seemed to be more severe than with the moderate doses. It was suggested that with a moderate dose, aggressive behavior was inhibited but the animals were able to adapt to each other. With a light dose, it was felt that aggressive behavior was not inhibited sufficiently, whereas with the high doses there was obviously suppression of fighting, but the animals may have been too sedated to adapt to each other (Symoens, 1969).

The approach of tranquilizing animals that are likely to fight when placed together may be used in instances in which there are just two animals involved. This would be the most useful in dogs, since while tranquilized, they are capable of settling the dominant-subordinate arrangement and incapable of a full-blown fight. Once the dominance hierarchy is clearly defined, the dose of tranquilizer is gradually reduced. The case of Bright and Zippy illustrates the use of a tranquilizer to ameliorate a dominance problem (Case 24–1).

Drug Choice in Animal Psychopharmacology

There is little detailed information available that can be used clinically as a basis in deciding which psychoactive drug to employ for a particular problem behavior; but choices must be made. In this section we will offer some information to guide such choices. The clinician must follow a type of guided trial and-error approach.

In attempting to alter a behavior using drug therapy, the best approach is to choose a representative from each biochemical group. With the exception of progestins, tranquilizers are virtually the only type of psychoactive drugs of value. Antidepressants, anticonvulsants, stimulants, and lithium carbonate have not yet proved useful in clinical animal behavior.

The dose should be high, but consistent with pharmacological safety limits. It might be that a moderate dose of a drug will not alter a behavior but a considerably higher dose will. It is known in human psychopharmacology, and observed in animals as well, that individuals with severe behavioral disorders often require a much higher dose of tranquilizer to alter behavior than normal individuals. Finally, dogs and cats, like

CASE 24–1

History. Fighting between Bright and Zippy has become so intense that their owners now keep them separated at all times. Bright, a Cocker Spaniel, has been with the family longer and is favored by the owners, but Zippy, a Doberman Pinscher, has recently grown much larger than Bright and seems to feel dominant. The owners feel that placing the dogs together would evoke a fight that would injure one or both of them.

Diagnosis. Dominance-Related Aggression

General Evaluation. The dogs will both be tranquilized and then reintroduced to each other. At the same time, it will be necessary to support the dominant position of Zippy.

SPECIFIC INSTRUCTIONS

1. Bright and Zippy will be tranquilized for two weeks with chlorpromazine (1.5/kg). Throughout this period the two dogs will be kept together to allow them to begin adjusting their dominant-subordinate interactions.
2. You are to begin favoring Zippy to support him as the dominant dog.
3. After the initial two week period, the tranquilizer dose will be lowered by one third per week.

people, show pronounced individual differences in reactions to different psychoactive drugs. A drug that gives gratifying results in one dog, may not affect another dog with the same apparent syndrome. Thus, even with more complete information on drugs, we must always be cognizant of individual reactions to drug types and dosages.

As part of a behavioral therapeutic program, one should always give the drug on a continuous basis with a dosage frequency of two or three times a day. The amount given per dosage is then gradually lowered as some form of behavioral therapy is continued.

Characteristics of Drugs

For practical purposes the drugs used in behavioral therapy are tranquilizers of two groups, phenothiazines and benzodiazepines. Of the phenothiazines, some of the more readily available for veterinary use are acetylpromazine (Acepromazine), chlorpromazine (Thorazine), and piperacetazine (Psymod). Of the benzodiazepines, diazepam (Valium) has pretty much replaced chlordiazepoxide (Librium) as the most potent and frequently used drug of the group. Until contrary information is available, we will assume that the basic behavioral effects of drugs are the same for all members of each group; the drugs may differ in mg/kg dosage for behavioral effectiveness, or in ease of evoking side effects, such as sedation or ataxia.

Phenothiazines. These are the drugs of choice in attempting to suppress types of aggressive behavior that do not reflect fear. Competition for dominance over the owner or fighting with other dogs are examples.

The usefulness of phenothiazines in treating aggressive behavior is supported by observations that chlorpromazine was effective in suppressing hostility in cats (Norton and deBeer, 1956) and induced aggressive behavior in laboratory rodents (Cook and Kelleher, 1963). These are also the indicated drugs for forms of hyperactivity. The phenothiazine drugs have a greater tendency to produce sedation than the benzodiazepines.

The phenothiazines enjoy wide use in pharmacological restraint and as preanesthetics for dogs and cats. This may reflect the tendency of the drugs to sedate as well as calm animals. Drugs of this group are more likely to suppress excitability than benzodiazepines (Hart, 1985c).

Benzodiazepines. These are the drugs of choice in attempting to alter fear reactions. Work on laboratory animals reveals that the benzodiazepines are the most effective for fear related behaviors (Cook and Kelleher, 1963; Yen, *et al.*, 1970) for medical examination or treatment. The taming effect of the benzodiazepines on fear-induced behavior in tigers, lions, dingos, monkeys, and other zoo animals has been quite impressive (David, 1966; Heuschele, 1961; Hines, 1960; Smits, 1964). Some striking experiences with the use of chlordiazepoxide are reported by Heuschele (1961). A vicious male European lynx that charged the front of the cage with teeth bared and snarling when an observer approached, became calm, licked an observer's finger, and allowed him to examine his paws when given 6 mg/kg orally. This dose also produced drowsiness and slight ataxia. A female dingo, normally very hostile to humans, became friendly, approaching an observer with her tail wagging and licked his hand with an oral dose of 3 mg/kg. This dose did not produce ataxia. A male dingo that normally cowered in the back of the cage also became friendly and approached observers with the same oral dose of chlordiazepoxide.

In limited observations on fear in dogs under laboratory conditions, diazepam has proven more effective than chlorpromazine in reducing fear (Hart, 1985c).

Diazepam has a tendency to increase excitability in otherwise relatively calm dogs. In dogs that are excitable, the drug, unlike thorazine or acepromazine, does not reduce excitability (Hart, 1985c).

USE OF DRUGS FOR SPECIFIC BEHAVIORAL SYNDROMES

In this final section, we will draw on the information available to suggest the most appropriate psychoactive drug for initial choice in treatment. It should be emphasized that the drug should be considered a tool or facilitator of behavioral therapy, not as the cure for a problem. It must also be remembered that, given individual animal differences in reactions to psychoactive drugs, if one drug does not produce the desired behavioral effect at a reasonable dosage, a drug from another group should be tried.

Cats

The information available for making recommendations is very limited. The suggestions are based on experimental work reported in the course of analyzing drugs and from the authors' clinical experience in using drugs for problem behavior in cats.

Fear. This includes the behavior of some animals to withdraw from people and hide under furniture on a continuous basis. The prognosis is most favorable for cases in which the withdrawn behavior is a recent occurrence. The drug of choice is diazepam (Valium) at 1 to 5 mg per adult cat two or three times a day. An example of the treatment of such a problem is presented in Case 24–2.

Aggression Toward Other Cats or People. The drug of choice is a phenothiazine such as thorazine, acepromazine or piperacetazine. If the aggression is toward another cat and is long-standing, the prognosis is quite guarded. Remember that cats are basically asocial animals and attempts to force them to live amicably with other cats in the same household, even with psychoactive drugs, go against the animal's basic nature.

Dogs

The basis for the recommendations for dogs are primarily from clinical case history reports and the authors' clinical experience.

Fear. This includes phobias that represent unhabituated reactions as well as phobias acquired through classical conditioning (see Chapter 22). This may be fear of loud noises or certain people, such as men, people in uniforms, or strangers. The drug of choice is diazepam (Valium) at 1 to 2 mg/kg two or three times per day. Treatment for such a problem is presented in Case 24–3.

Aggression Toward People and Other Dogs. The treatment of choice is a phenothiazine, such as thorazine, acepromazine or piperacetazine. Valium has also been effective in instances in which a phenothiazine has not. Two cases in which tranquilizers altered aggressive behavior are presented in Cases 24–4 and 24–5. Often these drugs are used in conjunction with some changes in behavioral interactions with the dog.

CASE 24–2

History. Pamela was always a shy cat, but recently she has totally withdrawn. She hides inside a cupboard whenever people, including the owner, are in the house and comes out only when no one is around.

Diagnosis. Fear-Related Withdrawal

General Evaluation. The cat will be tranquilized to reduce the fear reaction, while it is rewarded for being out among people.

SPECIFIC INSTRUCTIONS

1. Administer diazepam (1 to 2 mg) to Pamela three times a day.
2. Encourage Pamela to circulate in the house by relocating her feed dish and litter box away from the cupboard.
3. Continue tranquilizer therapy at full dosage for one week after she begins coming out of the cupboard. Over the next two weeks, reduce the treatment gradually by giving the drug twice a day and then once a day.

CASE 24–3

History. Shady, a Siberian Husky, jumped out of a moving van and was lost for the afternoon. Since then he has been fearful of strangers, at times urinating and defecating in their presence.

Diagnosis. Fear Reaction to People

General Evaluation. A benzodiazepam tranquilizer will be used to attempt to reduce Shady's fear to allow for desensitization sessions.

SPECIFIC INSTRUCTIONS

1. Administer diazepam (one 2 mg tablet) to Shady three times a day. If there is no favorable effect after 24 hours, double the dosage (two 2 mg tablets each time). If there is still no effect after 24 hours, double the dosage again (four 2 mg tablets each time). If there is then no effect, discontinue the drug for 2 days and begin administering the alternative, acetylpromazine.
2. If it is necessary to use acetylpromazine, give one 10 mg tablet three times a day. If there is no effect, double the dosage as above, until the dose is four 10 mg tablets three times a day.
3. If the drug evokes stumbling or excessive sleeping without behavioral effects, discontinue treatment.
4. Progress check visit in two weeks to evaluate further tranquilizer regimen and behavioral measures to be employed.

CASE 24–4

History. This German Shepherd, Rogue, is aggressive with the two boys in the family. He growls and shows piloerection. While he has not bitten anyone, he once lunged at the younger child.

Diagnosis. Dominance-Related Aggression

General Evaluation. Tranquilization with acetylpromazine will be used to assist the family members, especially the children, in asserting dominance over Rogue.

SPECIFIC INSTRUCTIONS

1. Administer the acetylpromazine (10 mg) tablet three times per day. If necessary double the dosage. Continue for two weeks.
2. Punish Rogue severely whenever he shows aggressive behavior. Use both verbal and physical punishment and then ignore him for at least one half hour. You might even try to evoke aggression if the tranquilizer is working.
3. Stop allowing Rogue to sleep in the parents' bedroom. Also, adults should not give any affection to Rogue except when the children are present.
4. The boys should begin mildly "bullying" Rogue, and also giving food treats after Rogue obeys several commands.
5. Reduce the tranquilizer dosage by one third over three weeks.

CASE 24–5

History. Spicy fights other dogs whenever they approach her. The owner has trained Spicy to obey commands, and these are effective except when a dog comes near Spicy.

Diagnosis. Aggression Toward Other Dogs

General Evaluation. A combination of tranquilization and affection withdrawal will be used to shift Spicy toward liking to be around other dogs.

SPECIFIC INSTRUCTIONS

1. Administer acetylpromazine to Spicy (one 10 mg tablet three times a day) for three weeks. Double the dosage if necessary.
2. Withdraw all affection from Spicy except when other dogs are present. Provide occasions for Spicy to be around other dogs frequently.
3. Stage visits of house guests who will bring a mild-mannered dog. Command that Spicy sit still while the visiting dog remains quietly at the other end of the room. As this process is repeated on successive days, gradually have the visiting dog move closer. During such times provide Spicy with abundant affection.
4. Reduce the dosage of acetylpromazine by one third each week if progress is sufficient at the end of three weeks.

References

Baron, A., Stewart, C.N., and Warren, J.M.: Patterns of social interaction in cats *(Felis domestica).* Behaviour, *11*:56, 1957.

Beach, F.A.: Coital behavior in dogs. III. Effects of early isolation on mating in males. Behaviour, *30*:218, 1968.

Beach, F.A., and Fowler, H.: Individual differences in the response of male rats to androgen. J. Comp. Psychol. *52*:50, 1959.

Beach, F.A., and Merara, M.: Coital behavior in dogs. V. Effects of estrogen and progesterone on mating and other forms of social behavior in the bitch. J. Comp. Physiol. Psych., *70*:1, 1970.

Beaver, B.: Veterinary Aspects of Feline Behavior. St. Louis, the C.V. Mosby Co., 1980.

Bland, K.P.: Tom-cat odour and other pheromones in feline reproduction. Vet. Sci. Commun., *3*:125, 1979.

Boulcott, S.R.: Feeding behaviour of adult dogs under conditions of hospitalization. Br. Vet. J., *123*:498, 1967.

Brodie, B.B.: Part IV. Difficulties in extrapolating data on metabolism of drugs from animal to man. Clin. Pharmacol. Ther., *3*:374, 1962.

Caro, T.M.: The effects of experience on the predatory patterns of cats. Behav. Neural Biol., *29*:1, 1980a.

Caro, T.M.: Effects of the mother, object play, and adult experience on predation in cats. Behav. Neural Biol., *29*:29, 1980b.

Chastain, C.B., Graham, C.L., and Nichols, C.E.: Adrenocortical suppression in cats given megestral acetate. Am. J. Vet. Res., *42*:2029, 1981.

Clemens, L.G.: The neurohormonal control of masculine sexual behavior. *In* Reproductive Behavior. Edited by W., Montagna and W.A. Sadler. New York, Plenum Press, 1974, pp. 23–53.

Cook, L., and Kelleher, R.T.: Effects of drugs on behavior. Ann. Rev. Pharmacol. *3*:205, 1963.

Corson, S.A., Corson, E.O., Kirilcuk, V., Kirilcuk, J., Knopp, W., and Arnold, L.E.: Differential effects of amphetamines on clinically relevant dog models of hyperkinesis and stereotypy: Relevance to Huntington's Chorea. *In* Advances in Neurology. Vol. 1. Edited by A. Barbeau. New York, Raven Press, 1972, pp. 681–697.

Corson, S.A., Corson, E.L'L., Gwynne, P.H., and Arnold, L.E.: Pet dogs as nonverbal communication links in hospital psychiatry. Compr. Psychiatry. *18*:61, 1977.

Daly, M.: Early stimulation of rodents: A critical review of present interpretations. Br. J. Psychol., *64*:435, 1973.

Damassa, D.A., Smith, E.R., Tennent, B., and Davidson, J.M.: The relationship between circulating testosterone levels and male sexual behavior in rats. Horm. Behav., *8*:275, 1977.

David, R.: Ro. 5-2807/BIO* (Valium) as a tranquilizer in zoo animals. Int. Zoo Yearbook, *6*:270, 1966.

Day, C., and Galef, B.: Pup cannibalism: One aspect of maternal behavior in golden hamsters. J. Comp. Physiol. Psychol., *91*:1179, 1977.

DeBoer, J.N.: Dominance relations in pairs of domestic cats. Behav. Processes, *2*:227, 1977.
Diakow, C.: Effects of genital desensitization on mating behavior and ovulation in the female cat. Physiol. Behav., 7:47, 1971.
Dunbar, I.: Dog Behavior: Why Dogs Do What They Do. Neptune, T.F.H. Publications, Inc., 1979.
Frank, D.W., Kirton, K.T., Murchison, T.E., Quinlon, W.J., Coleman, M.E., Gilbertson, T.J., Feenstra, E.S., and Kimball, F.A.: Mammary tumors and serum hormones in the bitch treated with medroxyprogesterone acetate or progesterone for four years. Fertil. Steril, *31*:340, 1979.
Garcia, J., Hankins, W.G., and Rusiniak, K.W.: Behavioral regulation of the milieu interne in man and rat. Science, *185*: 824, 1974.
Gentry, R., and Wade, G.N.: Androgenic control of food intake and body weight in male rats. J. Comp. Physiol. Psychol., *90*:18, 1976.
Gerber, H.A., and Sulman, F.G.: The effect of methyloestrenolone on oestrus, pseudopregnancy, vagrancy, satyriasis and squirting in dogs and cats. Vet. Rec., *76*:1089, 1964.
Gorski, R.A., Harlan, R.E., Jacobson, C.D., Shryne, J.E., and Southam, A.M.: Evidence for the existence of a sexually dimorphic nucleus in the preoptic area of the rat. J. Comp. Neurol., *193*:529, 1980.
Grunt, J.A., and Young, W.C.: Differential reactivity of individuals and the response of the male guinea pig to testosterone propionate. Endocrinology, *51*:237, 1952.
Gustavson, C.R., Garcia, J., Hankins, W.G., and Rusiniak, K.W.: Coyote predation control by aversive conditioning. Science *184*:581, 1974.
Hamburg, D.A., Moos, R.H., and Yalom, I.D.: Studies of distress in the menstrual cycle and the postpartum periods. *In* Endocrinology and Human Behavior. Edited by R.R. Michael. London, Oxford University Press, 1968.
Hamilton, M.: Prediction of clinical response from animal data: A need for theoretical models. *In* Animal Behavior and Drug Action. Edited by H. Steinberg. Boston, Little, Brown and Co., 1964, pp. 299-307.
Harding, C.F., and Feder, H.H.: Relation between individual differences in sexual behavior and plasma testosterone levels in the guinea pig. Endocrinology, *98*:1198, 1976.
Hart, B.L.: Role of prior experience in the effects of castration on sexual behavior of male dogs. J. Comp. Physiol. Psychol., *66*:719, 1968.
Hart, B.L.: The male reproductive system. *In* The Beagle as an Experimental Dog. Edited by A.C. Anderson. Ames, Iowa State University Press, 1970, pp. 296–312.
Hart, B.L.: Introduction. Feline Pract. *1(1)*:7, 45, 1971.
Hart, B.L.: The action of extrinsic penile muscles during copulation in the male dog. Anat. Rec., *173*:1, 1972.
Hart, B.L.: Environmental and hormonal influences on urine marking behavior in the adult male dog. Behav. Biol., *11*:167, 1974a.
Hart, B.L.: Gonadal androgen and sociosexual behavior of male mammals; A comparative analysis. Psychol. Bull., *81*:383, 1974b.
Hart, B.L.: Handling and restraint of the cat. Feline Pract., *5(2)*: 10, 1975a.
Hart, B.L.: Learning abilities in cats. Feline Pract., *5(5)*:10, 1975b.
Hart. B.L.: Behavioral effects of castration. Canine Pract., *3(3)*:10, 1976.
Hart, B.L.: Three disturbing behavioral disorders in dogs: Idiopathic viciousness, hyperkinesis and flank sucking. Canine Prac. *4(6)*:10, 1977.
Hart, B.L.: Breed-specific behavior. Feline Pract., *9(6)*:10, 1979a.
Hart, B.L.: Behavioral therapy with mousetraps. Feline Pract., *9(4)*:10, 1979b.
Hart, B.L.: Toilet training and problems with elimination. Canine Pract., *6(1)*:7, 1979c.
Hart, B.L.: Problems with objectionable sociosexual behavior of dogs and cats: Therapeutic use of castration and progestins. The Compendium on Continuing Education for the Small Animal Practitioner, *1*:461, 1979d.
Hart, B.L.: Evaluation of progestin therapy for behavioral problems. Feline Pract., *9(3)*:10, 1979e.
Hart, B.L.: Prescribing cats. Feline Pract., *10(1)*:8, 1980a.
Hart, B.L.: Starting from scratch: A new perspective on cat scratching. Feline Pract., *10(4)*:8, 1980b.
Hart, B.L.: Objectionable urine spraying and urine marking behavior in cats: Evaluation

of progestin treatment in gonadectomized males and females. J. Am. Vet. Med. Assoc., *177*:529, 1980c.

Hart, B.L.: Progestin therapy for aggressive behavior in male dogs. J. Am. Vet. Med. Assoc., *178*:1070, 1981a.

Hart, B.L.: Olfactory tractotomy to control objectionable urine spraying and urine marking in cats. J. Am. Vet. Med. Assoc., *179*:231, 1981b.

Hart, B.L.: Neurosurgery for behavioral problems. A curiosity or the new wave? *In* Symposium on Animal Behavior. Vol. 12. Edited by V.L. Voith and P. Borschelt. Philadelphia, Vet. Clin. North Am., W.B. Saunders Co., 1982.

Hart, B.L.: Role of testosterone secretion and penile reflexes in sexual behavior and sperm competition in male rats: A theoretical contribution. Physiol. Behav., *31*:823, 1983.

Hart, B.L.: Behavior of Domestic Animals. New York, W.H. Freeman & Co., 1985a.

Hart, B.L.: Urine spraying and marking in cats: Behavioral, pharmacological and surgical approaches. *In* Textbook of Small Animal Surgery. Edited by D.H. Slatter. Philadelphia, W.B. Saunders Co., 1985b.

Hart, B.L.: Behavioral indications for phenothiazine and benzodiazeprine tranquilizers in dogs. J. Am. Vet. Med. Assoc., 1985c, in press.

Hart, B.L., and Barrett, R.E.: Effects of castration on fighting, roaming, and urine spraying in adult male cats. J. Am. Vet. Assoc. *163*:290, 1973.

Hart, B.L., and Cooper, L.: Factors related to urine spraying and fighting in prepubertally gonadectomized male and female cats. J. Am. Vet. Med. Assoc., *184*:1255, 1984.

Hart, B.L., and Hart, L.A.: Selecting the best companion animal: Breed and gender specific behavioral profiles. *In* The Pet Connection: Its Influence On Our Health and Quality of Life. Edited by R.K. Anderson, B.L. Hart, and L.A. Hart. Minneapolis, University of Minnesota, 1984.

Hart, B.L., and Hart, L.A. Selecting pet dogs on the basis of cluster analysis of breed behavioral profiles and gender. J. Am. Vet. Med. Assoc., 1985, in press.

Hart, B.L., and Ladewig, J.: Serum testosterone of neonatal male and female dogs. Biol. Reprod., *21*:289, 1979b.

Hart, B.L., and Ladewig, J.: Accelerated and enhanced testosterone secretion in juvenile male dogs following medial preoptic-anterior hypothalamic lesions. Neuroendocrinology, *30*:20, 1980.

Hart, B.L., and Leedy, M.: Identification of source of urine stains in multi-cat households. J. Am. Vet. Med. Assoc., *180*:77, 1982.

Hart, B.L., and Miller, M.F.: Behavioral profiles of dog breeds: A quantitative approach. J. Am. Vet. Med. Assoc., 1985, in press.

Hart, B.L., Murray, S.R.J., Hahs, M., and Cruz, B.: Breed-specific behavioral profiles of dogs: Model for a quantitative analysis. *In* New Perspectives for Our Lives With Animal Companions. Edited by A. Katcher and A. Beck. Philadelphia, University of Pennsylvania Press, 1983.

Hart, B.L., and Peterson, D.M.: Penile hair rings in male cats may prevent mating. Lab. Anim. Sci., *21*:422, 1971.

Hart, B.L., and Voith, V.L.: Sexual behavior and breeding problems in cats. Feline Pract., *7(1)*:9, 1977.

Heuschele, W.P.: Chlordiazepoxide for calming zoo animals. J. Am. Vet. Med. Assoc., *136*:996, 1961.

Himwich, H.E.: Biochemical and neurophysiological action of psychoactive drugs. *In* Drugs and Behavior. Edited by L. Uhr and J.G. Miller. New York, John Wiley & Sons, Inc., 1960, pp. 41–85.

Himwich, H.E.: Locus of actions of psychotropic drugs in the brain. Folia Psychiatr. Neurol. Jpn., *19*:217, 1965.

Hines, L.R.: Methaminodiazeoxide (Librium): A psychotherapeutic drug. Curr. Ther. Res., *2*:227, 1960.

Holliday, T.A., Cunningham, J.G., and Gutnick, M.J.: Comparative clinical and electroencephalographic studies of canine epilepsy. Epilepsia, *11*:281, 1970.

Hopkins, S.G., Schubert, T.A., and Hart, B.L.: Castration of adult male dogs: Effects on roaming, aggression, urine marking, and mounting. J. Am. Vet. Med. Assoc., *168*:1108, 1976.

Houpt, K.A., Coren, B., Hintz, H.F., and Hilderbrant, J.E.: Effect of sex and reproductive

status on sucrose preference, food intake, and body weight of dogs. J. Am. Vet. Med. Assoc., *174*:1083, 1979.
Jarvick, M.J.: Drugs used in the treatment of psychiatric disorders. *In* The Pharmacological Basis of Therapeutics. Edited by L.S. Goodman and A. Gilman. New York, Macmillan Publishing Co., Inc., 1970, pp. 151–203.
Katcher, A.H., and Friedmann, E.: Potential health value of pet ownership. The Compendium of Continuing Education for the Small Animal Practitioner, *2*:117, 1980.
Larsson, K.: Individual differences in reactivity to androgen in male rats. Physiol. Behav., *1*:255, 1966.
Laundre, J.: The daytime behaviour of domestic cats in a free-roaming population. Anim. Behav., *25*:990, 1977.
Le Boeuf, B.J.: Copulatory and aggressive behavior in the prepuberally castrated dog. Horm. Behav., *1*:127, 1970.
Leigh, H., and Hofer, M.: Behavioral and physiological effects of littermate removal on the remaining single pup and mother during the preweaning period on rats. Psychosom. Med., *35*:497, 1973.
Leshner, A., and Collier, G.: The effects of gonadectomy on the sex differences in dietary self-selection patterns and carcass compositions of rats. Physiol. Behav., *11*:671, 1973.
Levine, S.: Infantile stimulation: A perspective. *In* Stimulation in Early Infancy. Edited by A. Ambrose. New York, Academic Press, Inc. 1969, pp. 3–19.
Leyhausen, P.: The communal organization of solitary mammals. Proc. Symp. Zool. Soc. Lond., *14*:259, 1965.
Leyhausen, P.: Cat Behavior. New York, Garland STPM Press, 1979.
Manocha, S.: Abortion and cannibalism in squirrel monkeys *(Saimiri sciureus)* associated with experimental protein deficiency during gestation. Lab. Anim. Sci., *26*:649, 1976.
Meier, G.W.: Infantile handling and development in Siamese kittens. J. Comp. Physiol. Psychol., *54*:284, 1961.
Mitchel, J.S., and Keesey, R.E.: The effects of lateral hypothalamic lesions and castration upon the body weight and composition of male rats. Behav. Biol., *11*:69, 1974.
Moyer, K.E.: Kinds of aggression and their physiological basis. Commun. Behav. Biol., Part A, *2*:65, 1968.
Mugford, R.A.: External influences on the feeding of carnivores. *In* The Chemical Senses and Nutrition. Edited by M.R. Kare and O. Maller. New York, Academic Press, Inc., 1977.
Nisbett, R.E.: Hunger, obesity, and the ventromedial hypothalamus. Physiol. Rev., *79*:433, 1972.
Norton, S., and deBeer, E.J.: Effects of drugs on the behavioral patterns of cats. Ann. NY Acad. Sci., *65*:249, 1956.
Pearson, O.P.: Carnivore-mouse predation: An example of its intensity and bioenergetics. J. Mammal., *45*:177, 1964.
Resnick, O.: The psychoactive properties of diphenylhydantoin: Experiences with prisoners and juvenile delinquents. Int. J. Neuropsychiatry, *3(2)*:S40, 1967.
Ross, S.: Some observations of the lair dwelling behavior of dogs. Behaviour, *12*:144, 1959.
Ross, S., and Ross, J.: Social facilitation of feeding behaviour in dogs. I. Group and solitary feeding. J. Genet. Psychol., *74*:97, 1949.
Schneirla, T.C., Rosenblatt, J.S., and Tobach, E.: Maternal behavior in the cat. *In* Maternal Behavior in Mammals. Edited by H.L. Rheingold. New York, John Wiley & Sons, Inc., 1963, pp. 122–168.
Scott, J.P., and Fuller, J.L.: Genetics and the Social Behavior of the Dog. Chicago, University of Chicago Press, 1965.
Seitz, P.: Infantile experience and adult behavior in animal subjects. Psychosomatic Med. *21*:353, 1959.
Simon, L.J.: The pet trap: Negative effects of pet ownership on families and individuals. *In* The Pet Connection: Its Influence on our Health and Quality of Life. Edited by R.K. Anderson, B.L. Hart, and L.A. Hart. Minneapolis, University of Minnesota, 1984.
Singh, P., and Hofer, M.: Oxytocin reinstates maternal olfactory cues for nipple orientation and attachment in rat pups. Physiol. Behav., *20*:385, 1978.
Smith, M.S., and McDonald, L.E.: Serum levels of luteinizing hormone and progesterone during the estrous cycle, pseudopregnancy, and pregnancy in the dog. Endocrinology, *94*:404, 1974.

Smits, G.M.: Some experiments and experiences with neuroleptic and hypnotic drugs on ungulates with special reference to librium. Tijdschr. Diergeneekd, *89(1)*:196, 1964.

Solarz, A.K.: Behavior. *In* The Beagle as an Experimental Dog. Edited by A.C. Anderson. Ames, Iowa State University Press, 1970.

Stern, J., and Levine, S.: Pituitary-adrenal activity in the postpartum rat in the absence of suckling stimulation. Horm. Behav., *3*:237, 1972.

Stabenfeldt, G.H.: Physiologic, pathologic and therapeutic roles of progestins in domestic animals. J. Am. Vet. Med. Assoc., *164*:311, 1974.

Symoens, J.: Prevention and cure of aggressiveness in pigs using the sedative azaperone. Vet. Rec., *85*:64, 1969.

Tanzer, H., and Lyons, N.: Your Pet Isn't Sick (He Just Wants You To Think So). New York, E.P. Dutton, 1977.

Tortora, D.F.: Safety training: The elimination of avoidance-motivated aggression in dogs. J. Exp. Psychol. Gen., *112*:176, 1983a.

Tortora, D.F.: Safety training: The elimination of avoidance-motivated aggression in dogs. Calif. Vet., *9*:15, 1983b.

Tuber, D.S., Hothersall, D., and Voith, V.L.: Animal clinical psychology: A modest proposal. Am. Psychol., *29*:762, 1974.

Vandenberg, S.G.: Behavioral methods of assessing neuroses and psychoses. *In* Drugs and Behavior. Edited by L. Uhr and J.G. Miller. New York, John Wiley & Sons Inc., 1960, pp. 463–500.

Van der Velden, N.A., DeWeerdt, C.J., Brooymans-Schallenberg, J.H.C., and Tielen, A.M.: An abnormal behavioral trait in Burmese mountain dogs. Tijdschr. Diergeneekd, *101*:403, 1976.

Voith, V.L.: Destructive behavior in the owner's absence. Part I. Canine Pract., *2(3)*:11, 1975a.

Voith, V.L.: Destructive behavior in the owner's absence. Part II. Canine Pract., *2(4)*:8, 1975b.

Voith, V.L.: Hyperactivity and hyperkinesis. Mod. Vet. Pract., *61*:787, 1980a.

Voith, V.L.: Fear-induced aggressive behavior. *In* Canine Behavior. Edited by B.L. Hart. Santa Barbara, Veterinary Practice Publishing Company, 1980b.

Voith, V.L.: Functional significance of pseudocyesis. Mod. Vet. Pract., *61*:75, 1980c.

Voith, V.L.: Attachment between people and their pets: Behavior problems of pets that arise from the relationship between pets and people. *In* Interrelations Between People and Pets. Edited by B. Fogle. Springfield, Charles C. Thomas, 1981.

Vollmer, P.J.: Electrical stimulation as an aid in training. Part 1. VM SAC, *Nov.*:1600, 1979a.

Vollmer, P.J.: Electrical stimulation. Part 2. Bark-training collars. VM SAC, *Dec.*:1737, 1979b.

Vollmer, P.J.: Electrical stimulation. Part 3. Conclusion. VM SAC, *Jan.*: 57, 1980.

vom Saal, F.S.: Variation in phenotype due to random intrauterine positioning of male and female fetuses in rodents. J. Reprod. Fertil., *62*:633, 1981.

Wade, G.N.: Sex hormones, regulatory behaviors, and body weight. *In* Advances in the Study of Behavior. Vol. 6. Edited by J.S. Rosenblatt, R.A. Hinde, E. Shaw, and C. Beer. New York, Academic Press, Inc., 1976, pp. 201–279.

Wilbur, R.H.: Pets, pet ownership and animal control. Social and psychological attitudes. Proc. Natl. Conf. Dog Cat. Cont. Am. Hum. Assoc., Denver, Colorado, February, 1976.

Yen, H.C.Y., Krop, S., Mendez, H.C., and Katz, M.H.: Effects of some psychoactive drugs on experimental "neurotic" (conflict induced) behavior in cats. Pharmacology, *3*:32, 1970.

Yerkes, R.W., and Bloomfield, D.: Do kittens instinctively kill mice? Psychol. Bull., 7:253, 1910.

Index

Page numbers in *italics* indicate figures. Page numbers followed by "t" indicate tables.